GE
BIOL

Blue Ridge Community and Technical College

Martinsburg, WV

Part I: General Chemistry, CHEM 127

Michael O'Donnell

Part II: Organic and Biological Chemistry, CHEM 128

Bruce Kowiatek

LABORATORY MANUAL FOR CHEM 127 & CHEM 128

endall Hunt
lishing company

MICHAEL J. O'DONNELL • BRUCE K. KOWIATEK

www.kendallhunt.com
Send all inquiries to:
4050 Westmark Drive
Dubuque, IA 52004-1840

ISBN: 978-1-5249-7698-9

Published in the United States of America

Contents

Periodic Table of the Elements

Atomic Number →	1	
	H	← Symbol
Name →	Hydrogen	
	1.008	← Atomic Weight

1 IA	2 IIA	3 IIIB	4 IVB	5 VB	6 VIB	7 VIIB	8 VIIIB	9 VIIIB	10 VIIIB	11 IB	12 IIB	13 IIIA	14 IVA	15 VA	16 VIA	17 VIIA	18 VIIIA
1 H Hydrogen 1.008																	2 He Helium 4.002602
3 Li Lithium 6.94	4 Be Beryllium 9.0121831											5 B Boron 10.81	6 C Carbon 12.011	7 N Nitrogen 14.007	8 O Oxygen 15.999	9 F Fluorine 18.998403163	10 Ne Neon 20.1797
11 Na Sodium 22.98976928	12 Mg Magnesium 24.305											13 Al Aluminium 26.9815385	14 Si Silicon 28.085	15 P Phosphorus 30.973761998	16 S Sulfur 32.06	17 Cl Chlorine 35.45	18 Ar Argon 39.948
19 K Potassium 39.0983	20 Ca Calcium 40.078	21 Sc Scandium 44.955908	22 Ti Titanium 47.867	23 V Vanadium 50.9415	24 Cr Chromium 51.9961	25 Mn Manganese 54.938044	26 Fe Iron 55.845	27 Co Cobalt 58.933194	28 Ni Nickel 58.6934	29 Cu Copper 63.546	30 Zn Zinc 65.38	31 Ga Gallium 69.723	32 Ge Germanium 72.630	33 As Arsenic 74.921595	34 Se Selenium 78.971	35 Br Bromine 79.904	36 Kr Krypton 83.798
37 Rb Rubidium 85.4678	38 Sr Strontium 87.62	39 Y Yttrium 88.90584	40 Zr Zirconium 91.224	41 Nb Niobium 92.90637	42 Mo Molybdenum 95.95	43 Tc Technetium (98)	44 Ru Ruthenium 101.07	45 Rh Rhodium 102.90550	46 Pd Palladium 106.42	47 Ag Silver 107.8682	48 Cd Cadmium 112.414	49 In Indium 114.818	50 Sn Tin 118.710	51 Sb Antimony 121.760	52 Te Tellurium 127.60	53 I Iodine 126.90447	54 Xe Xenon 131.293
55 Cs Caesium 132.90545196	56 Ba Barium 137.327	57 - 71 Lanthanoids	72 Hf Hafnium 178.49	73 Ta Tantalum 180.94788	74 W Tungsten 183.84	75 Re Rhenium 186.207	76 Os Osmium 190.23	77 Ir Iridium 192.217	78 Pt Platinum 195.084	79 Au Gold 196.966569	80 Hg Mercury 200.592	81 Tl Thallium 204.38	82 Pb Lead 207.2	83 Bi Bismuth 208.98040	84 Po Polonium (209)	85 At Astatine (210)	86 Rn Radon (222)
87 Fr Francium (223)	88 Ra Radium (226)	89 - 103 Actinoids	104 Rf Rutherfordium (267)	105 Db Dubnium (268)	106 Sg Seaborgium (269)	107 Bh Bohrium (270)	108 Hs Hassium (269)	109 Mt Meitnerium (278)	110 Ds Darmstadtium (281)	111 Rg Roentgenium (282)	112 Cn Copernicium (285)	113 Nh Nihonium (286)	114 Fl Flerovium (289)	115 Mc Moscovium (289)	116 Lv Livermorium (293)	117 Ts Tennessine (294)	118 Og Oganesson (294)

57 La Lanthanum 138.90547	58 Ce Cerium 140.116	59 Pr Praseodymium 140.90766	60 Nd Neodymium 144.242	61 Pm Promethium (145)	62 Sm Samarium 150.36	63 Eu Europium 151.964	64 Gd Gadolinium 157.25	65 Tb Terbium 158.92535	66 Dy Dysprosium 162.500	67 Ho Holmium 164.93033	68 Er Erbium 167.259	69 Tm Thulium 168.93422	70 Yb Ytterbium 173.045	71 Lu Lutetium 174.9668
89 Ac Actinium (227)	90 Th Thorium 232.0377	91 Pa Protactinium 231.03588	92 U Uranium 238.02891	93 Np Neptunium (237)	94 Pu Plutonium (244)	95 Am Americium (243)	96 Cm Curium (247)	97 Bk Berkelium (247)	98 Cf Californium (251)	99 Es Einsteinium (252)	100 Fm Fermium (257)	101 Md Mendelevium (258)	102 No Nobelium (259)	103 Lr Lawrencium (266)

LAB 1

Measurement

Measurement—whether we realize it or not, we are constantly measuring things. Typically, this is not formalized wherein a specific instrument is used to take a measurement with certain units. Instead, what we do is mentally measure as we drift through our day. For instance, as you drive your early morning commute, you may not look at the speedometer to determine your velocity at a given point in the drive, however, you are comparing your speed to other cars around you and to the fixed markers along the route. You could not with certainty say at what speed you are moving, yet you can say comparatively that you move at the same speed as other cars. This type of measurement is qualitative measurement. We establish the quality of one set of actions relative to another set.

There are two separate ways to measure. The first was mentioned above, qualitative measurement. The second is quantitative measurement. Where the first type of measurement compares two or more qualities, the second uses a specific instrument to quantify the measurement. Notice the difference in the terms: qualitative—quality; quantitative—quantity. We begin our observations in science with a qualitative measurement but then must move to making a similar quantitative measurement.

Taking a sick child's temperature is an example of this. We place our hand on the child's forehead to determine whether the child is warmer than our hand (qualitative). If the child feels warmer than our hand, we then place a thermometer in the child's mouth and take the temperature (quantitative). We move from a comparison to an actual measurement that contains a number with matching units. It is this quantitative measurement that we must have when collecting data in a scientific experiment.

Some further examples of measurement:

Jane is taller than Sarah versus Jane is 5 feet 6 inches and Sarah is 5 feet 3 inches.

Today is hotter than yesterday versus today's temperature is 30°C versus yesterday's temperature was 23°C.

Though we have a number and unit of measurement when we use a measuring instrument, this does not mean that uncertainty does not exist in this quantitative measurement. In fact, a quantitative measure consists of a series of known values with an added digit that is an approximation or guess. A device used for making a measurement is only as accurate as the manufacturer has made it. Look at a ruler or meter stick as an example. Each is used for linear measurements. Depending on how the ruler was made, one might see tick marks every decimeter, or centimeter, or millimeter. Often, when we use a given ruler, we find that that which is measured falls between two different tick marks. It is at this point that we must guess how far along the ruler the measurement should go and this becomes the final number in the measurement. The following diagram shows an arrow between two rulers with different sets of tick marks. Each ruler gives a different measurement that indicates the accuracy of the tool used and therefore the measurement recorded. For the top ruler, a measurement of the arrow would be 10.9 cm, whereas using the lower ruler would return a measurement of 10.85 cm. However, both measurements have uncertainty since the final digit of each is an estimate of where the arrow is.

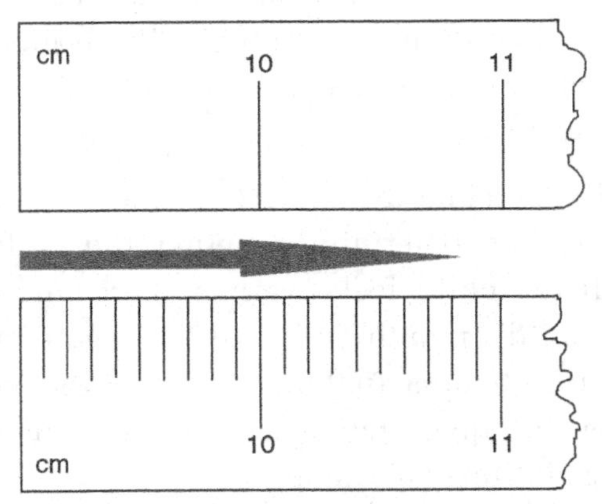

Source: Michael O'Donnell

In this lab, you will take various quantitative measurements as a way to familiarize yourself with not only how measurements are made but also the different types of tools used to take measurements.

Learning Outcomes

1. Differentiate between qualitative and quantitative measurement
2. Explain how quantitative measurement is better suited to scientific inquiry
3. Use various measurement instruments (balance, scales, graduated cylinder, ruler)
4. Convert units within the metric system
5. Calculate experimental error as a means of determining accuracy and precision in measurement

Materials

1. Various measuring tools: digital balance, triple beam balance, ruler, graduated cylinder
2. Pencil
3. Calculator
4. Lab report

Procedures

There are several items placed on the lab table. The following instructions list the type of measurement to make for each of these items. Record the measurement in the data table and then convert the measured unit into the asked for unit.

1. Metal cylinder: use the ruler and/or caliper to measure the length of the metal cylinder on the lab table and record this length in the data table in units of millimeter. Then measure the diameter of the base of the cylinder with the caliper and record this measurement in the data table. Use the following formula to calculate the volume of the cylinder and record the volume in the

data table. Finally, measure the cylinder using the displacement method and record in the data table.

Length (Height) (cm)	Diameter (cm)	Calculated Volume (cm^3) Vol = $\pi r^2 h$	Volume by Displacement (cm^3)

2. Mass of pennies: use the digital scale to measure and record the mass of the pennies on the table. Note that there are two stacks. One stack contains pennies minted before 1983. The other stack contains pennies minted after 1983. Keep the stacks separate and mass all of the pennies in a stack at the same time. Calculate the mass per penny and record in the data table.

Total Mass Pre-1983 Pennies (g)	Mass Per Penny (g)	Total Mass Post-1983 Pennies (g)	Mass Per Penny (g)

3. Fill the beaker on the lab table to the top mark on the side with tap water. Carefully pour the water from the beaker into the graduated cylinder. (CAUTION: you may have more water in the beaker than can be measured in the graduated cylinder. In this case, pour as much water into the graduated cylinder as you are able, dump it out, then pour in the rest of the tap water from the beaker.) Record the volume read from the beaker and the volume from the graduated cylinder in the data table. Explain any difference.

Volume as Read on Side of Beaker (mL)	Volume Measured with Graduated Cylinder (mL)

LAB 2

Density

The density of a substance is defined as the mass divided by the volume: $d = m/v$. Density is a physical property of a substance that does not depend on the amount of material present and is therefore called an **intrinsic property**. This physical property, like melting point (MP) and freezing point (FP), helps one to discover what an unknown substance is. In this experiment, you will find the density of water and the density of a metal by directly measuring mass and volume.

Materials: Dropper, water, metal sample, balance, 50-mL graduated cylinder

A. Procedure (For Metal Solid)

1. Obtain your sample and find the mass to the nearest **0.01 g**. Record the mass on the **data table** as "mass of metal sample."
2. Put enough water into your 50-mL graduated cylinder so that when you add the sample, it will be completely submerged, but the water level will not go beyond your graduations. Between 20 and 30 mL will be the best. Read the volume and estimate to the nearest **0.1 mL**. There is no room for error, so be very accurate. Record this volume on the **data table** as "initial volume of water."
3. Place your sample into the water in the graduated cylinder. Do this very carefully so that no water splashes out. Tip the cylinder so that the sample slides in slowly.
4. Make sure that the sample is completely submerged and no air is trapped in the sample. Read the volume to the nearest **0.1 mL** and record it in the **data table** as the "volume of water and metal sample."

Data Table

a. Mass of metal sample: __________ g

b. Initial volume of water: __________ mL

c. Volume of water and metal sample: __________ mL

Calculations

5. Find the volume of the sample. Remember, when something is submerged in water, it displaces water equal to its volume. You have recorded the volume of the water, and the volume of the water plus the sample. This leaves you with a simple subtraction (show calculation below).

Volume of metal sample = ________ mL

6. Now calculate the density of your sample; the sample's mass divided by the sample's volume. The unit for your density will be g/mL since no units cancel (show calculation below).

$$\text{Density} = \frac{\text{Mass (g)}}{\text{Volume (mL)}}$$

Density of metal = ________ g/mL

Use the following formula for experimental error (the theoretical value will be provided based on the chosen metal):

$$\text{Experimental error} = \frac{|\text{Theoretical value} - \text{Expiremental value}|}{\text{Theoretical value}} \times 100$$

B. Procedure (For Density of Water)

1. Find the mass of an empty, dry 50-mL graduated cylinder to the nearest 0.01 g and record the mass on the **data table**.

2. Add exactly **10.0 mL** of water to the cylinder. **Remember**, the **bottom** of the meniscus should just be touching the 10.0 mL line. **[Hint: Add water up to about the 9-mL mark and use a dropper to reach the 10.0-mL mark.]**
3. Find the mass of the cylinder and 10 mL of water to the nearest 0.01 g. Record the mass on the **data table.**
4. Repeat steps 2 and 3 with **30.0 mL** of water.
5. Repeat steps 2 and 3 with **50.0 mL** of water.

Data Table

Mass of empty graduated cylinder: ____________ g

Mass of graduated cylinder and 10.0 mL of water: ________ g; Mass of 10.0 mL of water: ______ g

Mass of graduated cylinder and 30.0 mL of water: ________ g; Mass of 30.0 mL of water: ______ g

Mass of graduated cylinder and 50.0 mL of water: ________ g; Mass of 50.0 mL of water: ______ g

A method to determine the density of a substance is graphically. Below is a graph on which you will plot your data for volume and mass of water from the above procedure. Once you have plotted the three points of data (volume vs. mass for each) draw a best-fit straight line to represent the trend of the data. Finally, pick two coordinate points to determine the slope of this line and compare it to your calculated density of water. Slope compares the rise of the line (moving up the *y* axis) to the run (moving along the *x* axis).

$$\text{Slope} = \frac{y_2 - y_1}{x_2 - x_1}$$

Show your slope calculations and show a comparison to the calculated density:

Slope: ____________ g/mL Calculated density: ___________ g/mL

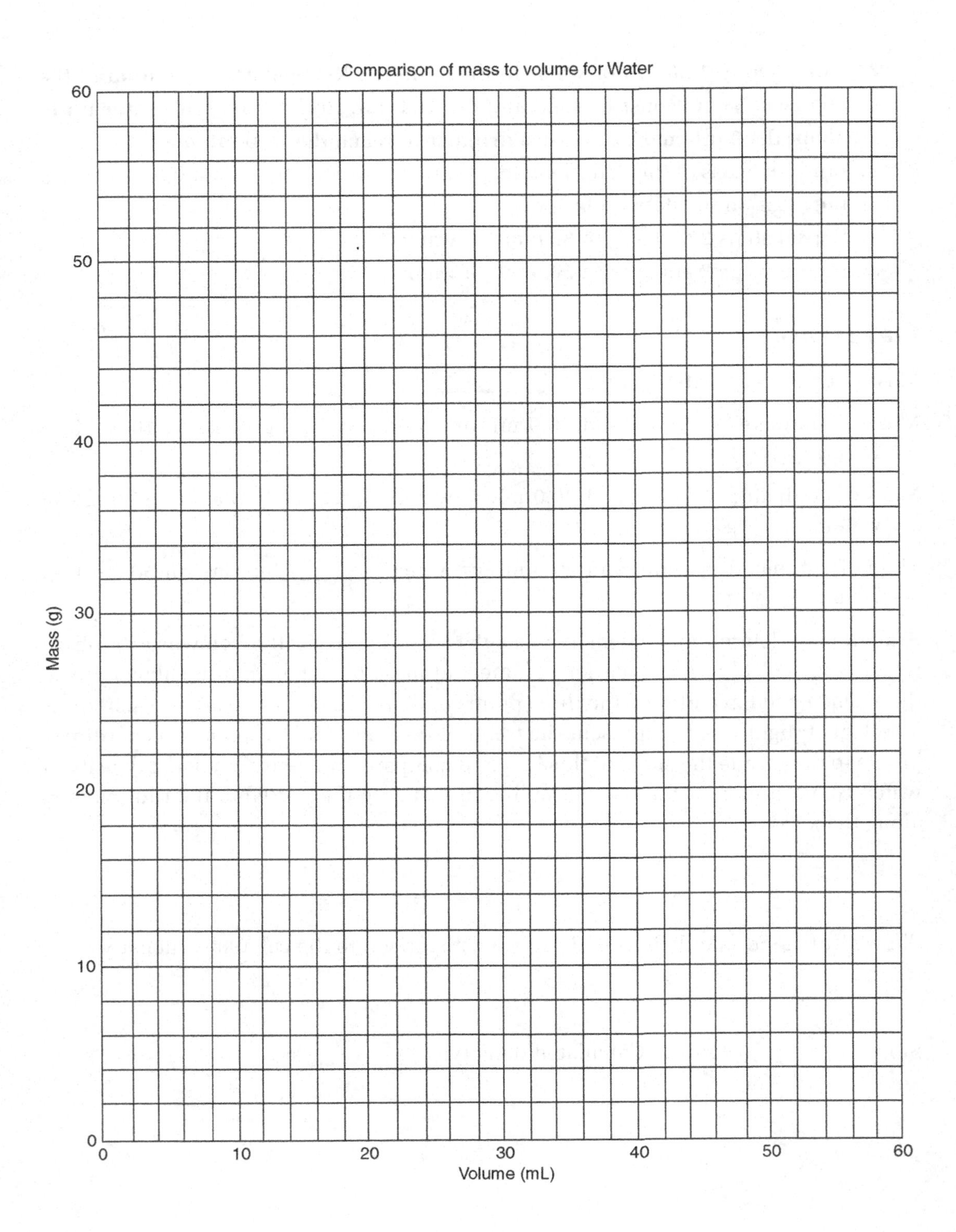
Comparison of mass to volume for Water
Mass (g)
Volume (mL)
0
10
20
30
40
50
60

Calculations (Show All Calculations in Spaces Provided)

6. Find the mass of the 10.0 mL sample of water. This is the **(mass of the cylinder and 10.0 mL sample) _ (the mass of the cylinder).**

   Mass of the 10.0 mL sample = ________ g

7. Calculate the density of the 10.0-mL sample. Remember:

$$\text{Density} = \frac{\text{Mass (g)}}{\text{Volume (mL)}}$$

   The unit for density will be **g/mL.**

   Density for 10.0 mL sample = ________ g/mL.

8. Now repeat steps 6 and 7 for the 30.0-mL and 50.0-mL samples.

   Mass of the 30.0-mL sample = ________ g

   Density of 30.0-mL sample = ________ g/mL

   Mass of the 50.0 mL sample = ________ g

   Density of 50.0 mL sample = ________ g/mL

   Average density of water = ____________g/mL

   The density of water is **1.0 g/mL.**

   How do your three answers compare to the accepted value?

   Calculate the average of your three densities for water above and use that average to determine your experimental error in the space below.

   Use the following formula for experimental error:

$$\text{Experimental error} = \frac{|\text{Theoretical value} - \text{Expiremental value}|}{\text{Theoretical value}} \times 100$$

Questions and Problems

1. You measure the mass and volume of an unknown solid that is shiny and has a golden luster. The following are you results: mass = 780 g, volume = 162.5 cm^3

a. Calculate the density of this material (show your work below).

b. Compare your calculated density to that of gold (Au), 19.3 g/cm^3, and determine if your sample is gold or something else. Explain why it is or is not gold (Au).

c. If it is not gold, explain how to determine what the substance is based on density

2. Using the following information, calculate the density of an unknown liquid and determine what that liquid is (circle the liquid in the given list):

   a. Mass of graduated cylinder is 50.67 g, mass of graduated cylinder and liquid is 54.82 g, the volume of the liquid is 3.28 mL (show your work)

Fluid	**Density** (g/mL) at 20°C and 1 atm unless noted
water (4°C)	0.99997
gasoline	0.66–0.69
ethyl alcohol	0.791
turpentine	0.85
mercury	13.55

LAB 3

Specific Heat

Specific heat, or specific heat capacity, C_p, is defined as the amount of energy needed to raise the temperature of 1 g of any substance by 1°C. The units of specific heat in SI are Joules per gram degree Celsius, J/g°C. Perhaps a better way to think about specific heat is to think about how much energy a substance can absorb without its temperature changing. Think of a teakettle on the stove. You turn the stove top on and wait for the water to come to a boil; the time spent waiting is longer for that kettle of water than it would be for a kettle of alcohol since the specific heat of water is 4.184 J/g°C and that of alcohol is 2.46 J/g°C.

Specific heat is, in essence, a conversion factor that allows the chemist to calculate energy needed to elevate the temperature of a given mass of a substance, or to calculate the specific heat using energy, mass, and change of temperature thus determining an unknown substance by identifying the specific heat of that substance. Below is the equation that is used to when working with temperature, mass, energy, and specific heat.

$$Q = m \times C_p \times \Delta T$$

Q is heat in Joules

m is mass in grams

C_p is specific heat in J/g°C

ΔT is temperature in °C

Specific heat of metals (as shown in Table 3.1) varies with the atomic mass of a given metal. As part of this lab, you will plot these points on a graph to determine this relationship.

Table 3.1 Specific Heat Capacity Table

	Specific Heat (J/g°C)	Atomic Mass (amu)
Aluminum	0.900	26.98
Zinc	0.39	65.39
Copper	0.385	63.55
Tin	0.21	118.71
Lead	0.160	207.20

Source: Michael O'Donnell

In this experiment, you will use a Styrofoam cup calorimeter as a means to collect data of temperature change and mass, and, knowing the specific heat of water, calculate the specific heat of an unknown metal. We assume that all of the energy absorbed by the heated metal completely transfers to the water in the calorimeter. Therefore, the Q_{water} equals the Q_{metal}. You will then compare this calculated value to the accepted value of your sample metal using the experimental error equation:

$$\text{Experimental error (\%)} = \frac{|\text{Accepted value} - \text{Experimental value}|}{\text{Accepted value}} \times 100$$

Finally, you will explain sources of error in your procedure based on the experimental error calculated.

Prelab Study Questions

1. Which of the following equations is used to determine the energy needed to change the temperature of water?

   a. $\vec{F} = m\vec{a}$

   b. $Q = m \times C_p \times \Delta T$

   c. $a^2 + b^2 = c^2$

   d. $\mathrm{A} = \pi \mathrm{r}^2$

2. Define specific heat, C_p?
3. You measure the starting temperature of 200 g of water and determine it is 35°C. If the water is heated to 40°C, calculate the energy needed to heat up this water (show all work).
4. If the energy in the above example was provided by a heated metal sample with a mass of 150 g, calculate the specific heat of this sample. The change in temperature was 71.5°C (show all work).
5. Using Table 3.1, determine the material of the sample.

Materials: 2 Styrofoam cups, 2 thermometers, metal sample, 400-mL beaker, hot plate, thongs.

Procedure

1. Weigh a dry metal sample and record the mass in the data table along with the letter code stamped on the bottom the sample.
2. Fill the 400-mL beaker approximately two-thirds full with tap water and place the beaker on the hot plate. Turn on the hot plate and place the metal sample in the beaker of water. Allow the water to come to boiling and boil with the metal for at least 10 minutes.
3. Nest the two cups to each other and weigh them. Record the mass in the data table.
4. Fill the cup with distilled water to just below the halfway mark, reweigh, and record the mass of the cups and distilled water in the data table.
5. Using one of the thermometers, measure the temperature of the water in the calorimeter to the nearest 0.5°C and record in the data table as the initial temperature of calorimeter water.
6. Use the second thermometer to measure the temperature of the boiling water and metal and record in the data table to the nearest 0.5°C. Assume the temperature of the metal is the same as the temperature of the boiling water. This is the initial temperature of the metal.
7. After you record the temperature of the boiling water, and the temperature of the calorimeter, and the metal has been in the boiling water bath for 10 minutes, you are ready to transfer the metal into the calorimeter. You must

do this quickly so that you reduce the time the metal is exposed to the air in the room. Use the thongs to grab the metal sample from the boiling water bath; then carefully place it in the calorimeter so that you do not splash water from the calorimeter.

8. Immediately watch the thermometer still in the calorimeter and record the highest temperature reached after placing the hot metal in the calorimeter. Record this to the nearest 0.5°C as the final temperature of water and metal in the data table.
9. If time allows, you will repeat these steps with a second metal sample.

Data Table

	Trial 1	Trial 2
Mass of metal sample (and symbol)	g	g
Mass of calorimeter	g	g
Mass of calorimeter and water	g	g
Mass of water	g	g
Initial temperature of water in calorimeter	°C	°C
Initial temperature of metal in boiling water bath	°C	°C
Final temperature of water and metal sample	°C	°C
Change in temperature of water (ΔT_{water}) in calorimeter	°C	°C
Change in temperature of metal sample(ΔT_{metal})	°C	°C
Specific heat (C_p) of water	4.184 J/g°C	4.184 J/g°C
Heat (Q_{water}) gained by water, show calculation		
Heat (Q_{metal}) lost by metal		

	Trial 1	Trial 2
Specific heat (C_p) of metal sample, show calculation		
Theoretical specific heat of metal (from table)		
Experimental error, show calculation (actual value given in above table)		

Questions and Problems

1. Why should you transfer the metal sample from the boiling water bath to the calorimeter as quickly as possible?
2. How much energy is required to raise the temperature of 25 g of water by 25°C?
3. A metal sample with a mass of 32 g is heated to 96°C, and then transferred to a calorimeter with 100 g of water at 25°C. If the final temperature of the calorimeter with the metal is 30°C, what is the specific heat of the metal sample?
4. List the sources of error in this experiment, and explain what you can do to minimize these sources of error.
5. Use the graph below to represent specific heat as a function of atomic mass for the data in Table 3.1. First plot each sample on the graphs, then draw a best-fit curve for the data (do not just connect the dots). Adjustable curves are available along with rulers if needed.

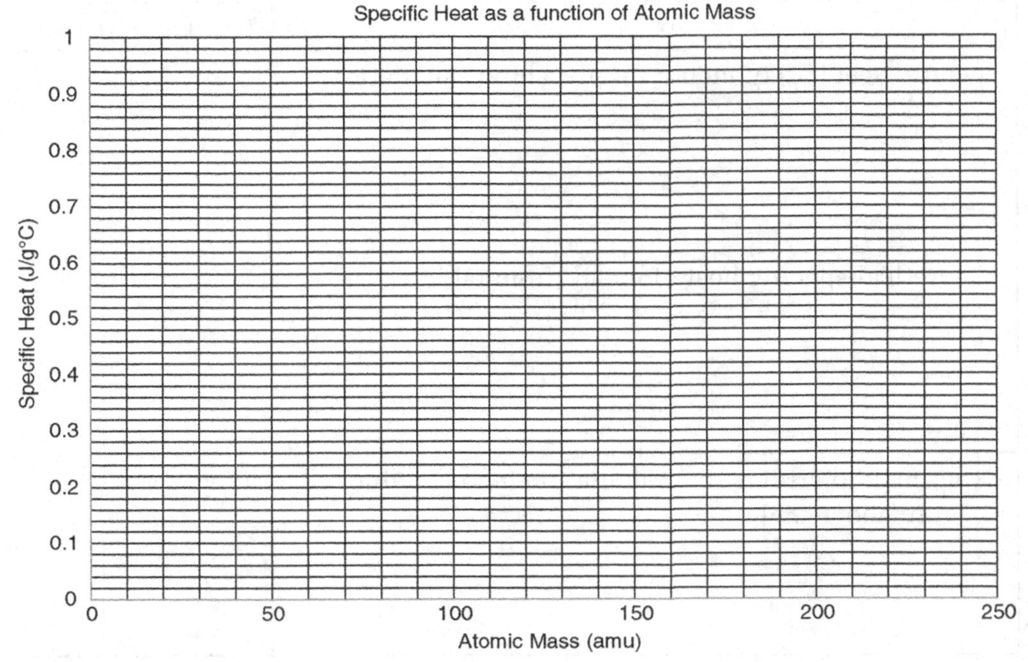

Source: Michael O'Donnell

6. Summarize the relationship of specific heat to the atomic mass of these metals.
7. Since the graph above represents a trend in specific heat when compared to atomic mass, you can use it to determine the atomic mass of a substance for which you have experimentally determined the specific heat. Given the information below, determine the atomic mass for the two substances:
   a. Experimental specific heat of 0.54 J/g°C ______________ amu
   b. Experimental specific heat of 0.18 J/g°C ______________ amu

LAB 4

Periodic Table

Dmitri Mendeleev, Lothar Meyer, and several other chemists felt that a table that shows the relationship between all the known elements would benefit everyone concerned. Mendeleev gets credit for the first periodic table as he made predictions that places left empty by him would be filled with a yet to be discovered element. His prediction was confirmed when Paul Emile discovered the element gallium, Ga, in 1875.

The periodic table of the elements, in essence, is a compact way to see all of the known elements and the relationships between these elements. For Mendeleev, it was an important device for his students to better understand the elements. Mendeleev used known atomic masses of the elements as well as combination properties between elements to arrange the table. Today, chemists have further refined the periodic table based on knowledge of quantum theory.

To use the periodic table effectively, one needs to be familiar with all of the information found on the periodic table, as well as the arrangement of the elements. As the title implies, the periodic table of the elements arranges the elements. The most basic information is the element name and symbol. Elements are the building blocks of all the chemical compounds one will use throughout this semester. Each card on the table contains the name of the element as well as the symbol for that element. Notice that some element names do not necessarily correspond to the symbol for that element. As an example look at element number 11 on the periodic table, sodium. The symbol for sodium is Na, and this symbol derives from the archaic name of this element, natrium. The symbol for most elements is based on the name of the given element and consists of a capital letter and a lower case letter if needed. (For instance, carbon is C and chlorine is Cl.) The periodic table is arranged in family groups (columns) as well as periods (rows) of all known elements. This provides the user with a simple way to point to any given element. For example, find the element in group 3A, period 6 and one sees that this is the element thallium, Tl. Over the years, certain groups have gained a special name. Column 1A is the alkali metals, 2A is the alkaline earth metals, 7A is the halogens, 8A is the noble gases, and the large block of elements in the center are the transitional metals. Yet, the table gives so much more information than this. For instance, as one follows along a period of elements, one discovers that atomic mass and atomic number increase. Other information on most periodic tables includes element name, and element symbol, atomic number, atomic mass, electronegativity, electron configuration.

Each element card contains two important numbers: the smaller of these is the atomic number which indicates the number of protons that element contains. The periodic table is arranged by increasing atomic number. The larger of these numbers is atomic mass, sometimes referred to as mass number. This is the total number of protons and neutrons contained in the nucleus of an element, which makes up the mass of an atom of that element. Notice that atomic masses are not whole numbers as one expects knowing it is total number of protons and neutrons. Most elements have several isotopes that contain varying numbers of neutrons (an element always has the same number of protons). This gives us atomic masses based on an average atomic mass for all isotopes of an element.

In this lab, students will explore some of the basics of the periodic table of the elements. It is hoped that by the end of lab, students are comfortable with using this tool throughout the semester.

Prelab Study Questions

1. In which row or column are the alkali metals located? In which are the halogens located?
2. From the following list of element symbols, circle those symbols that represent transitional elements and underline those elements that represent halogens:

   Mg Cu Br Ag Ni Cl Fe F

3. Complete the following table of elements and symbols:

Name of Element	Symbol	Name of Element	Symbol
Potassium			Na
Sulfur			P
Nitrogen			Fe
Magnesium			Cl
Copper			Ag

Materials: Lab manual, periodic table, display of elements, colored pencils/pens.

Procedure

1. Write the symbol and the atomic number for each listed element
2. Observe the elements on display; describe the color and luster (shininess) of each element
3. Identify from your observations each element as a metal (M), metalloid (ML), or a nonmetal (NM)

Element	Symbol	Atomic Number	Color	Luster	Metal/Nonmetal/ Metalloid
Aluminum					
Carbon					
Copper					
Iron					
Magnesium					
Sulfur					
Tin					
Zinc					

4. On the partial periodic table below, write the symbols and atomic numbers of the elements observed above, placing them in the proper place
5. Write the period numbers on the left side of the table
6. Outline the section that contains the transition elements
7. Draw a heavy line to separate metals from nonmetals

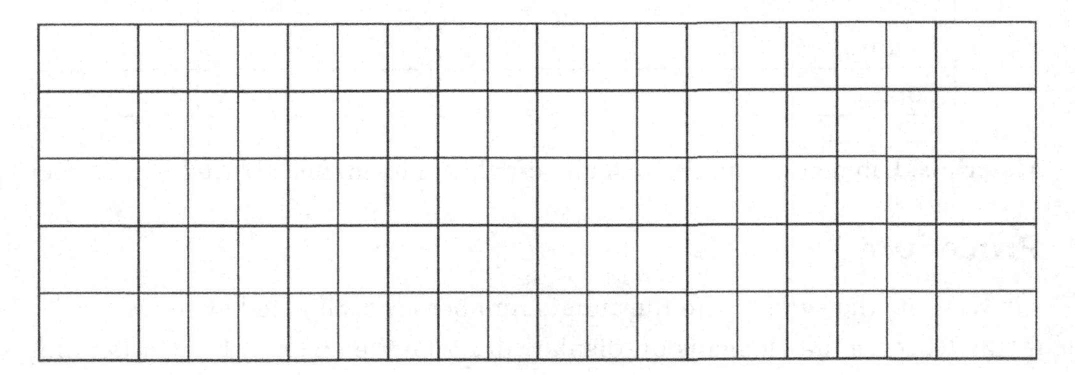

8. Using a periodic table, fill in the following chart with the information that is missing:

Name of Element	Symbol of Element	Atomic Number	Atomic Mass	Protons	Neutrons	Electrons
		3				3
Chromium			52			
	O			8		
		79	197			
				20	20	

Questions and Problems

1. From their positions on the periodic table, categorize each of the following elements as a metal, nonmetal, or a metalloid: (place an X in the appropriate box).

Element	Metal	Metalloid	Nonmetal
Sodium, Na			
Sulfur, S			
Copper, Cu			
Fluorine, F			
Iron, Fe			
Carbon, C			
Arsenic, As			
Calcium, Ca			

2. Give the symbol for each of the following elements:
   a. The noble gas in period 2 ______
   b. The halogen in period 2 ______
   c. Alkali metal in period 3 ______
   d. Halogen in period 3 ______
   e. Alkali metal in period 4 ______
   f. Metalloid in period 2 ______
3. A neutral atom has mass number of 80 and has 45 neutrons. Write the atomic symbol for this element.

4. An atom has four more protons and four more electrons than the above element. What is the element symbol and name?

LAB 5

Electron Configuration and Periodicity

As we move further into an examination of the periodic table and its arrangement of the elements, we notice a connection within the elements and how the electrons are configured for each element. From the previous lab exercise, you were able to determine a connection between the arrangement of elements on the periodic table regarding an element's atomic mass and number. In the early twentieth century, with the discoveries of the subatomic particles (protons, neutrons, and electrons), physicists and chemists alike desired to develop a model of how these particles are arranged in any given element. Niels Bohr proposed one of the earliest models of atomic structure referred to as a planetary model since he showed electrons orbiting the nucleus as the planets orbit our Sun. This model worked fine for the first several elements, however, scientists noticed that once you move beyond the element calcium ($^{40}_{20}Ca$) this simplified arrangement begins to degrade. In time, the quantum model of the atom was developed, relying on the quantum equation to determine the location of electrons. The quantum numbers, which make up the quantum equation, represent the best guess for the location of any given electron. There are four quantum numbers designated as follows: n = the principal quantum number and indicates the main energy level of the electron; l = the subshell where an electron is found; m = the orbital within the subshell; s = describes the spin of an electron. Within the context of this course, we shall simplify how we describe an electron's location by using electron configurations.

An electron configuration allows one to determine the location of all electrons for a given element. More importantly, it allows one to focus on the valence (outermost) electrons of an element. The valance electrons are those that are important when determining how elements combine to form compounds. In chemistry, we use an alphanumeric system for electron location. An integer, 1 through 7, is used in place of the principal quantum number; a letter (s, p, d, f . . .) is used to designate the subshell for an electron's location; and finally, a superscript integer (1 through 14) is used to determine the orbital in which an electron is located. The graphic below puts together a means to locate the electrons for a given element.

Principal Energy Level	s	p	p	p	d	d	d	d	d	f	f	f	f	f	f	f
4	↓↑	↓↑	↓↑	↓↑	↓↑	↓↑	↓↑	↓↑	↓↑	↓↑	↓↑	↓↑	↓↑	↓↑	↓↑	↓↑
3	↓↑	↓↑	↓↑	↓↑	↓↑	↓↑	↓↑	↓↑	↓↑							
2	↓↑	↓↑	↓↑	↓↑												
1	↓↑															

Source: Michael O'Donnell

The principal energy level corresponds to the row in which the element is located. The s, p, d, f denotes where along that row the element is found: s orbitals are the first 2 columns, p orbitals are the last 6 columns, d orbitals are the 10 transitional columns in the middle of the periodic table, and f orbitals are the 2 transitional rows shown outside of the periodic table. Each box represents a position where up to two electrons are located. Two electrons can collocate since the spin of each electron is in an opposite direction. The arrow in the boxes represent the electrons. In this manner, you can account for all electrons that an element has, starting with filling in the lowest possible energy level and continuing to fill in boxes until you have accounted for all the electrons a given element has. However, this would be cumbersome to fill in for quite a few elements, especially the larger elements of the periodic table that contain many electrons. Therefore, there is a simpler method of electron configuration used. It is a list of the electrons beginning again with the first energy level (row) and listing each subsequent row all on the same line. The principal energy level is the number, followed by the letter of the orbital that contains electrons, followed by a superscript that denotes how many electrons are in the orbital. Looking at the figure above, notice that any s orbital for any energy level can only contain, at most, 2 electrons; likewise p orbitals can contain up to 6 electrons; a d orbital can contain up to 10 electrons; and an f orbital up to 14 electrons. For example, aluminum, $^{27}_{13}\text{Al}$, contains 13 electrons (atomic number = 13 protons, therefore, 13 electrons). Therefore, we can show the electron configuration for aluminum that indicates all of the electrons and the location of the electrons around the nucleus of aluminum.

$$1s^2 2s^2 2p^6 3s^2 3p^1$$

This is easier than filling in a figure of boxes and relates the same information. If one adds up all of the superscript numbers in the electron configuration, it should equal the atomic number for that element.

A further aid in electron configuration is the chart to the left. As the valence shells for the large elements fill up, it is not a linear pattern for which shell is next to fill. There is a bit of a crossover when looking at the d- and f orbitals. The diagram to the left shows what happens in a general way with electron filling. One works across a row for the main energy level until one hits the diagonal arrow. At this point, one works down the diagonal until you reach an s orbital. Then, it is a process of working along the diagonals to the next energy level.

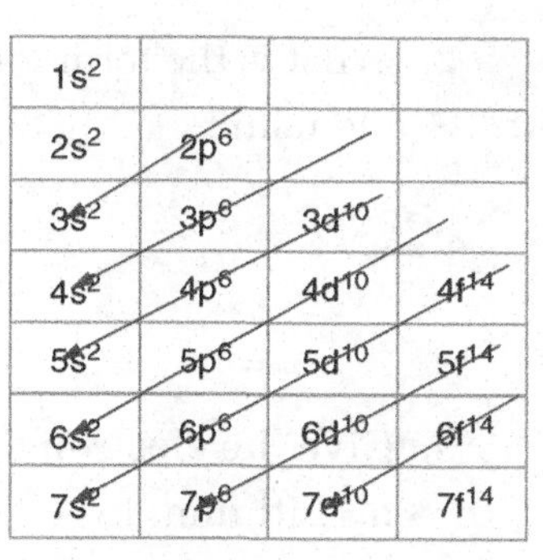

Source: Michael O'Donnell

Electron configurations provide a way to show location of electrons in a ground state. Flame tests help to identify an element as the electrons absorb energy, jump to a higher energy level, and then fall back to a lower energy level when the electron releases the energy. Each element has a signature color flame when heated. Part of this lab will look at flame tests and how a chemist may use this method to identify an unknown element. The following table will help with identifying the elements.

Element	Flame Color
Lithium, Li	Red
Sodium, Na	Persistent yellow-orange flame
Potassium, K	Lilac or pink
Calcium, Ca	Orange red
Strontium, Sr	Red
Barium, Ba	Pale green
Copper, Cu	Blue-green (sometimes white flashes)

Prelab Study Questions

1. What color would a street lamp give off if the gas inside the lamp were based on potassium, K? Explain why this is the color of the lamp.

2. What is the term for the outermost energy level containing electrons for a given element?

3. Give the electron configurations for the following elements:
   a. Lithium, Li

   b. Sodium, Na

   c. Potassium, K

4. Why do these elements all end with the same sublevel?

Materials: Saturated wood splints of various salts, source for an open flame (propane torch, alcohol lamp), unknown saturated wood splints, periodic table, chart of elements showing atomic number and atomic radius.

Procedures

A. Flame Tests
   1. Given various wood splints that are soaking in a saturated salt solution, hold each in an open flame and record the color of the flame produced from the salt.

2. Repeat the process with those wood splints of unknown salts and try to identify the unknown salt solution.

Solution	Element	Flame Color
Calcium chloride, $CaCl_2$	Calcium, Ca	
Potassium chloride, KCl	Potassium, K	
Barium chloride, $BaCl_2$	Barium, Ba	
Strontium chloride, $SrCl_2$	Strontium	
Sodium chloride, NaCl	Sodium, Na	
Unknown number 1		
Unknown number 2		
Unknown number 3		

B. Electron Configurations

Atom	Electron Configuration	Number of Valence Electrons	Group Number
Lithium, Li			
Carbon, C			
Neon, Ne			
Magnesium, Mg			
Sulfur, S			
Iron, Fe			
Potassium, K			
Copper, Cu			
Strontium, Sr			
Iodine, I			
Uranium, U			

C. Atomic Radius

Use the following table to fill in the accompanying graph for atomic radius as a function of atomic number.

Period	Element	Atomic Number	Atomic Radius (Picometer)
First	Hydrogen, H	1	37
	Helium, He	2	50
Second	Lithium, Li	3	152
	Beryllium, Be	4	111
	Boron, B	5	88
	Carbon, C	6	77
	Nitrogen	7	70
	Oxygen	8	66
	Fluorine	9	64
	Neon, Ne	10	70
Third	Sodium, Na	11	186
	Magnesium, Mg	12	160
	Aluminum, Al	13	143
	Silicon, Si	14	117
	Phosphorus, P	15	110
	Sulfur, S	16	104
	Chlorine, Cl	17	99
	Argon, Ar	18	94
Fourth	Potassium, K	19	231
	Calcium, Ca	20	197
	Scandium, Sc	21	160
	Titanium, Ti	22	150
	Vanadium, V	23	135
	Chromium, Cr	24	125
	Manganese, Mn	25	125

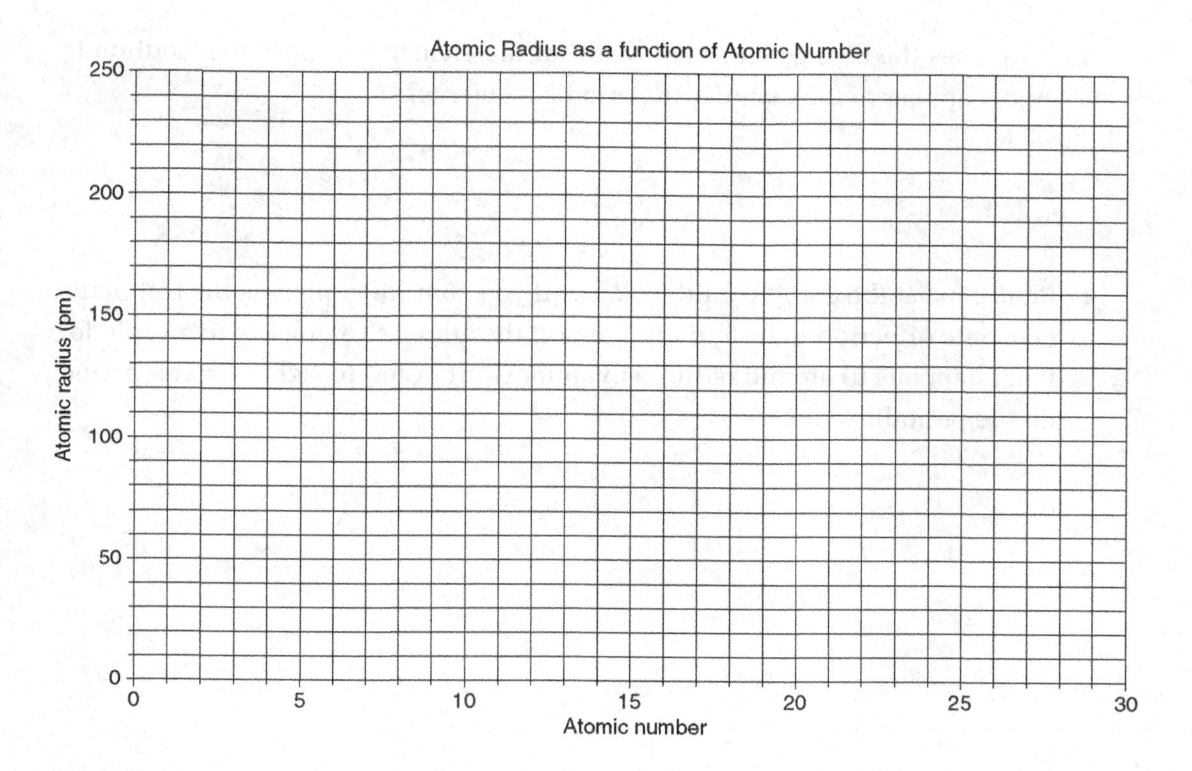

Questions and Problems

1. You are watching a fourth of July fireworks show and you notice the many colors overhead as the fireworks explode. Provide an explanation, based on your knowledge from the flame tests, for the various colors you see.

2. Using the graph above, describe the change in the atomic radii for the elements in period 2, lithium to neon.

3. Why does the change for atomic radii of the elements in period 3, sodium to argon, appear to look similar to period 2 elements?

4. Predict what the graph would look like if you continued placing the rest of the elements of period 4, iron to krypton, on the graph. Provide a general rule for what happens to atomic radius as you move from left to right across a period on the periodic table.

LAB 6

Compounds: Ionic versus Covalent

Now that you have explored the dynamics of the periodic table and determined the arrangement of elements based on electron configuration, we will explore how elements form compounds. All compounds are combinations of two or more elements. However, the way in which the elements combine varies in two ways. The electron configurations you did in the previous lab allow you to focus on just the valence electrons of any given element. This is important since it is the valence electrons that determine how an element combines with others to form a compound. Each type of compound has its own unique properties, physical and chemical, depending on the type of bond it forms. These unique properties help to identify that compound.

The first of these we call ionic compounds. Ionic compounds form when valence electrons from one element are transferred to the valence shell of another element. Since you now understand how to determine which electrons are valence electrons (especially when using the shorthand configuration), it is simple to predict if the elements will transfer electrons and form an ionic bond. A rule of thumb is elements contained in the groups 1, 2, or 13 of the periodic table will lose electrons and those elements in groups 15, 16, and 17 will gain electrons. Cations are those elements that lose valence electrons and anions are those elements that gain electrons. Cations are positively charged, with a charge equal to the number of electrons lost; anions are negatively charged with a charge equal to the number of electrons added. Typically, ionic compounds are a combination of a metal ion with a nonmetal ion.

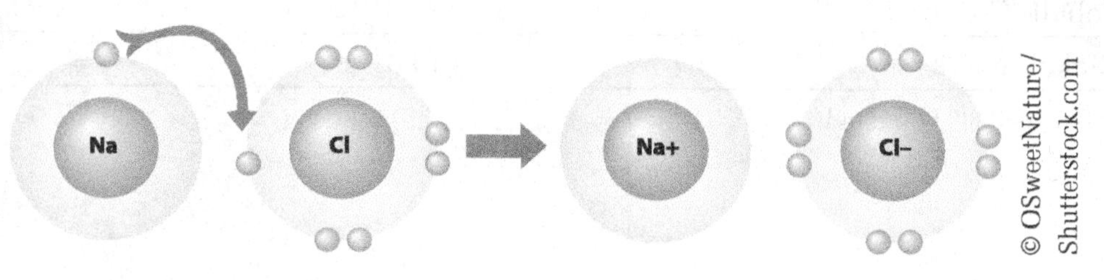

In this example, sodium, 11 on the periodic table, transfers its single valence electron to chlorine, 17 on the periodic table. By losing its lone electron, sodium becomes a +1 cation; by gaining a single electron, chlorine completes its valence shell and becomes a −1 cation. As stated above, the groups 1, 2, and 13 (3A) typically lose the same number of electrons as the group number (1, 2, and 3). Then there are the transition metals. In this area of the periodic table, electrons are filling shells that are overlapping in a sense. Thus, these elements have multiple ways that electrons are removed. Elements such as iron, Fe, and copper, Cu have multiple ionic charges that depend on how many valence electrons are lost. Let us look at the electron configuration of iron and determine the ionic charges it can have.

Fe $[Ar]4s^2 3d^6$

The electrons that are typically lost first are those in the highest energy level. For iron that would be the two electrons in the fourth energy level giving a cation of Fe^{2+}. However, the 3-d sublevel has one electron that iron could lose, which would provide stability by dropping the ion to a half-filled sublevel. This would give a cation of Fe^{3+}. The following chart shows several of the multivalenced cations for the transition metals we see most often. Whether only one valence cation or multivalenced cation, the goal is to form a compound that has electric neutrality between the cation and the anion.

Transition Element	Possible Valence Charges
Iron, Fe	2^+ and 3^+
Copper, Cu	1^+ and 2^+
Tin, Sn	2^+ and 4^+
Lead, Pb	2^+ and 4^+
Chromium, Cr	2^+ and 3^+ and 6^+
Cobalt, Co	2^+ and 3^+
Nickel, Ni	2^+ and 4^+

Source: Michael O'Donnell

The last piece of the puzzle for ionic compounds is the polyatomic ion. As the term suggests, they contain two or more elements bonded together (typically a covalent bond) but do not have electric neutrality. Polyatomic ions should be considered as a single unit when working with writing formulas and working with compounds. The following is a chart of common polyatomic ions:

Valence Charge	Polyatomic Ion
1^+	Ammonia, NH_4^{1+}
1^-	Nitrate, NO_3^{1-}
	Nitrite, NO_2^{1-}
	Chlorate, ClO_3^{1-}
	Chlorite, ClO_2^{1-}
	Hypochlorite, ClO^{1-}
	Hydroxide, OH^{1-}
2^-	Carbonate, CO_3^{2-}
	Sulfate, SO_4^{2-}
	Sulfite, SO_3^{2-}
3^-	Phosphate, PO_4^{3-}
	Phosphite, PO_3^{3-}

Source: Michael O'Donnell

On the other hand, covalent compounds, or molecules, contain two or more non-metal elements bonded together by sharing electrons. Neither element can remove the valence electrons needed to provide electron stability. So, a sharing of electrons is needed. The number of shared electrons in the bond connecting the elements varies depending on the number of valence electrons each element has. For example, fluorine, F, has nine electrons, seven of these electrons are valence. By adding one more electron, fluorine would complete its valence shell and take on a noble gas configuration. This is the reason why fluorine exists as one of the seven diatomic molecules (H_2, O_2, N_2, F_2, Cl_2, Br_2, I_2); this bonding provides electric stability for these elements.

An example of covalent sharing is the compound carbon dioxide. A carbon, four valence electrons, shares with two oxygens, six valence electrons.

The figure to the left shows the covalent bonding between the central carbon atom and the oxygen atoms. Carbon shares two of its four electrons with each oxygen and oxygen shares its two valence electrons with the carbon. In each bond, the shared electrons complete the octet noble gas configuration for the two elements involved in the bonding. Another rule of thumb is that the number of shared electrons between two covalently bonded elements must provide an octet to both elements. The exception to this rule is hydrogen that can only have two valence electrons.

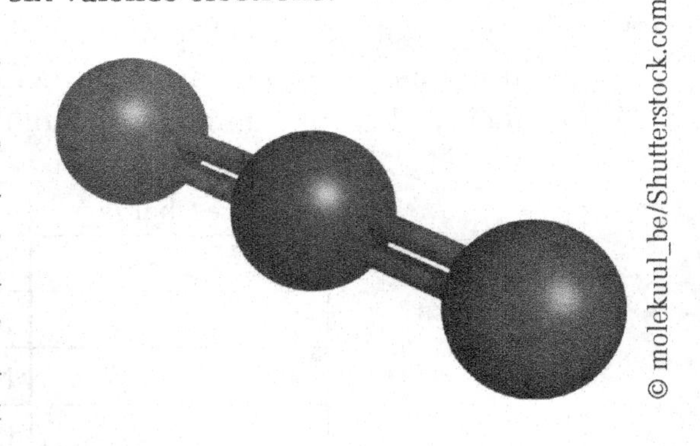

When forming covalent bonds, the molecules will assume a particular shape according to the Valence Shell Electron Pair Repulsion theory (VSEPR). The shared electrons exert a repulsive force on all other electron pairs around the central atom. The shape that the molecule has will then influence whether the molecule is nonpolar (as the above diagram of CO_2) or the molecule is polar, as is seen in the water molecule, H_2O shown here.

Source: Michael J O'Donnell; ACD/ChemSketch

Polarity influences how molecules interact with other compounds. A polar molecule will have intermolecular attractions with other polar molecules or with ionic compounds. For example, water is a solvent for most ionic compounds such as sodium chloride, NaCl. It is also a solvent for many other polar molecules like most alcohols. However, since water forms a polar molecule, it does not act as a solvent when mixed with a nonpolar molecule like gasoline or oil. The following table shows how molecular shape is affected by VSEPR and how shape leads to polarity.

Number of Electron Pairs	Electron Pair Geometries: 0 Lone Pair	1 Lone Pair	2 Lone Pairs
2	180° X—E—X Linear		
3	120° Trigonal planar	<120° Bent or angular	
4	109° Tetrahedral	<109° Trigonal pyramid	<<109° Bent or angular

Source: Modified from: https://files.mtstatic.com/site_4334/57957/0?Expires=1549730751&Signature =NUS8YflUztzX7y61G1TSkPrubbkxq0IPj1pm1dCuwGopyNHgffXrFQoQPZxaRGftG6oa~8cwxOil1m12 VqIlb3T4c9R7Sa-zkCtPHZSlmbqjSa3Z33PvkuFcuVyPQstQnIz42HXK8WZI1wS1EGluQP3rnIX2UleBU4 CRCnPHTmo_&Key-Pair-Id=APKAJ5Y6AV4GI7A555NA

Prelab Study Questions

1. Define valence shell electrons and describe the location in any given atom.

2. Describe the formation of a cation, positively charge ion, and an anion, negatively charge ion. Why do elements form an ion?

3. What are polyatomic ions?

4. Give the name for each of the following compounds:

a. NaCl

b. Fe_2O_3

c. $NaNO_3$

d. CO_2

5. What shape would each of the following compounds have?

a. Carbon dioxide

b. Sulfur trioxide

c. Ammonia

d. Water

e. Methane

Part A Formation of Ions

Complete the following chart showing the formation of these ions (you can use short-hand configuration):

Element	Atomic Number	Electron Configuration	Electron Dot Form	Lose or Gain e^-	Ion Electron Configuration	Ionic Charge	Ionic Symbol	Ionic Name
Sodium	11	$1s^2 2s^2 2p^6 3s^2$		Lose 1 e^-	$1s^2 2s^2 2p^6$	1^+	Na^+	Sodium
Nitrogen								
Aluminum								
Chlorine								
Calcium								
Oxygen								

Part B Ionic Compounds and Formulas

1. Physical properties

Sodium chloride, NaCl

a. Appearance

b. Density

c. Melting point

2. Complete the following table for these ionic compounds:

Name	Cation	Anion	Formula
Lithium iodide	Li^{+}	I^{-}	LiI
Aluminum oxide			
Caladium sulfide			
Magnesium bromide			
Potassium nitride			
Sodium fluoride			

3. Complete the following table for these ionic compounds:

Formula	Cation	Anion	Name
K_2S			
BaF_2			
MgO			
$AlCl_3$			
Mg_3P_2			
Li_3N			

Part C Multivalence Cations

1. Physical properties

   Iron (III) chloride, $FeCl_3$

   a. Appearance

   b. Density

   c. Melting point

2. Complete the following table for these multivalence ions:

Name	Cation	Anion	Formula
Iron (III) chloride			
Iron (II) oxide			
Copper (I) sulfide			
Copper (II) nitride			
Zinc oxide			
Silver sulfide			

Formula	Cation	Anion	Name
Cu_2S			
Fe_2O_3			
$CuCl_2$			
FeS			
Ag_2O			

Part D Polyatomic Ions

1. Physical properties

   Potassium carbonate, K_2CO_3

   a. Appearance

   b. Density

   c. Melting point

2. Complete the following tables containing polyatomic ions.

Name	Cation	Anion	Formula
Potassium carbonate			
Sodium nitrate			
Calcium hydroxide			
Lithium phosphate			
Magnesium sulfate			

Formula	Cation	Anion	Name
$CaSO_4$			
$Al(NO_3)_3$			
Na_2CO_3			
$MgSO_3$			
$Cu(OH)_2$			
$Mg_3(PO_4)_2$			

Part E Molecular Compounds

1. Physical properties

Glucose, $C_6H_{12}O_6$

a. Appearance

b. Density

c. Melting point

2. Complete the following tables for molecular compounds

Name	Formula	Name	Formula
Dinitrogen pentoxide		Dinitrogen trisulfide	
Siicon tetrachloride		Oxygen difluoride	
Phosphorus tribromide		Iodine heptafluoride	
	ClF_2		SF_6
	CS_2		N_2O_3
	PCl_5		SeF_6

LAB 7

Chemical Reactions

In this lab, we will explore how we determine whether a substance has gone through a chemical change. We discussed in lecture the difference between physical and chemical properties and physical and chemical changes. A physical property and by association a physical change is usually a change of phase or state. For example, an ice cube is water in the solid state. Through heating, we can change its state to liquid. For us to return to the ice, we merely remove the energy that was added. The substance we began with has not been altered. Physical changes are easily reversed. By contrast, a chemical change results in a new substance or substances forming. Let us use the water from the above example, but in this instance, we will add that to a bit of iron (say an old bicycle) and let it sit outside for a few days. You will notice that rust begins forming on the bike! This is a chemical change. The iron has combined with oxygen present in the atmosphere and the water to form iron oxide, Fe_2O_3. For this change to reverse, another chemical change is required.

But, how do we know a chemical change has taken place? In the above example, we see the formation of the rust (Fe_2O_3). Not always are these changes so dramatic. However, there are three observations we can make to support the idea of a chemical change. The first is formation of a precipitate. When aqueous (a substance dissolved in water) solutions of some compounds are combined, a chemical change occurs and one of the products is no longer soluble in water. By use of a solubility table, we can predict if a precipitate will form prior to mixing the solutions together. The second observation is the formation of a gas. Certain compounds, when mixed together, produce a gas that bubbles out of the solution. An example of this is placing zinc metal into a solution of hydrochloric acid, HCl. The zinc pulls the hydrogen from the chloride ion and the hydrogen gas forms bubbles rising in the solution. The third observation is a change in temperature. Certain chemical reactions either produce energy as the reaction proceeds (exothermic) or require energy input for the reaction to proceed (endothermic). The above example of zinc and hydrochloric acid is exothermic, the reaction vessel becomes noticeably hot. The cold packs used by athletes when injured is an example of an endothermic reaction of two chemicals.

Once we know how to identify whether a chemical reaction has occurred, we need a method to show this reaction on paper. Thus, chemists developed a shorthand for showing chemical reactions using the formulas for the compounds involved in the reaction. Basically, one shows the elements or compounds one begins with (the reactants) on the left side of the chemical equation with an arrow pointing to the right side of the equation (the products). The law of conservation of mass dictates that we cannot have more (or less) atoms on the product side as we begin with on the reactant side, therefore, we must balance the chemical equation.

Balancing chemical equations is a process where one places coefficients in front of a compound or elements formula to indicate that more of these are needed. The coefficient is a multiplier for everything in the formula that follows until some other sign is in place. Take the chemical reaction for the formation of water:

$$H_2 + O_2 \rightarrow H_2O$$

Notice that we have two hydrogen atoms and two oxygen atoms to begin on the reactant side, however, when we shift to the product side, we have only one oxygen atom present. This does not follow the law of conservation of atoms. We must end with two oxygen atoms so a coefficient of two is placed in front of H_2O like so:

$$H_2 + O_2 \rightarrow 2H_2O$$

There are now two hydrogen atoms in the reactants, but now there are four hydrogen atoms in the product, so, a coefficient of two is placed in front of the H_2 atom in the reactants.

$$2H_2 + O_2 \rightarrow 2H_2O$$

Now the reaction is balanced showing we begin with four hydrogen atoms, two oxygen atoms, and end up with two water molecules (each molecule contains two hydrogen atoms and one oxygen atom).

In observing chemical reactions, there are five common categories into which we place a given reaction. Those five categories along with a generalized chemical equation are summarized in the following table:

Type of Reaction	Description	General Form
Combination	Combining elements or compounds into one product	A + B → AB
Decomposition	Breaking a compound down into simpler compounds or elements	AB → A + B
Single replacement	An element replaces one of the elements in a compound	A + BC → B + AC
Double replacement	Elements of two different compounds switch places	AB + CD → AD + CB
Combustion	Shows the burning of a hydrocarbon to form CO_2 and H_2O	Hydrocarbon + O_2 → CO_2 + H_2O

Prelab Study Questions

1. In each of the following state whether there is a physical change or a chemical change.
   a. Bicycle rusting
   b. Snow melting
   c. Cake baking
   d. Egg cooking
   e. Steam rising from the road after the rain
2. List the three observations that support a chemical change and give an example of each.

3. What is the difference between a combustion reaction and a double displacement reaction?

4. Balance the following reactions and identify the type.

a.	___Al + ___Fe_2O_3 → ___Al_2O_3 + ___Fe	type:
b.	___$KClO_3$ → ___KCl + ___O_2	type:
c.	___ Li + ___Cl_2 → ___ LiCl	type:
d.	___NaCl + ___$AgNO_3$ → ___AgCl + ___$NaNO_3$	type:

Part A: Magnesium and Oxygen

Materials: Magnesium ribbon (2–3 cm long), tongs, propane burner.

1. Obtain the magnesium ribbon and record its appearance.
2. Using the tongs, hold the ribbon in the flame of the propane burner until it ingites. CAUTION: when the magnesium ribbon ignites DO NOT look directly at the flame! Record your observations about the reaction and the product of the magnesium reaction.
3. Write and balance a reaction equation for the magnesium and identify the type of reaction.

Initial appearance of Mg:

Evidence of a chemical reaction:

Balance chemical equation: ___Mg + ___ O_2 → ___MgO

Type of chemical reaction:

Part B: Zinc and Copper (II) Sulfate

Materials: Test tubes, test tube rack, 0.1M $CuSO_4$, Zinc.

1. Place two dropperfuls of copper (II) sulfate into a small test tube. Describe the appearance of the solution.

2. Obtain a small piece of zinc metal and describe its appearance.
3. Add the zinc metal to the test tube containing the copper (II) sulfate solution. Place the test tube in the test tube rack and note any changes in the zinc metal or the solution. Leave in the test tube rack and observe again after 15 minutes.
4. Balance the chemical reaction and identify the type of reaction.

	$CuSO_4$ Appearance	Zn Appearance	Evidence of a Reaction
Initial			
After 15 minutes			

___Zn + ___$CuSO_4$ → ___Cu + ___ $ZnSO_4$ type of reaction:

Part C: Reactions of Metals with HCl

Materials: Test tubes, test tube rack, small pieces of zinc, copper and magnesium, 3M HCl.

1. Obtain small pieces of zinc, copper, and magnesium. Describe the appearance of each of these metals.
2. Place two dropperfuls of 3M HCl into each of three test tubes. Carefully add each metal piece to a separate test tube. Observe and record any reactions in the test tubes, listing evidence that a reaction has taken place.
3. Balance the chemical equations for each metal that reacted with the HCl. If there was no reaction, then write no reaction next to that equation.
4. Identify the type of reaction you observed.

Metal	Appearance of Metal	Evidence of Chemical Reaction
Zn		
Cu		
Mg		

___Zn + ___HCl → type:

___Cu + ___HCl → type:

___Mg + ___HCl → type:

Part D: Reactions of Ionic Compounds

Materials: four test tubes, test tube rack, 0.1M solutions of $CaCl_2$, Na_3PO_4, $FeCl_3$, and KSCN.

1. Mark each test tube so that you know which test tube holds which solution.
2. Place two dropperfuls of each solution into the four test tubes. Record the appearance of each solution.
3. Pour the contents of the test tube that contains the Na_3PO_4 into the test tube that contains the $CaCl_2$. Record any evidence of a chemical reaction.
4. Pour the contents of the test tube that contains the KSCN into the test tube that contains the $FeCl_3$. Record any evidence of a chemical reaction.
5. Dispose of the contents of the test tubes as instructed.

Reactants	Appearance of Solutions	Evidence of a Reaction
$CaCl_2$		
Na_3PO_4		
$FeCl_3$		
KSCN		

___ $CaCl_2$ + ___ Na_3PO_4 → type:

___ $FeCl_3$ + ___ KSCN → type:

Part E: Sodium Carbonate and HCl

Materials: Test tube, test tube rack, 3M HCl, Na_2CO_3, matches.

1. Place three dropperfuls of 3M HCl in a test tube. Record the appearance of HCl. Obtain a small amount (pea size, or the very end of a scoopula) Na_2CO_3 and record its appearance.
2. Carefully add the Na_2CO_3 to the test tube that contains the HCl. Record any evidence of a chemical reaction.
3. Light a match or wood splint and carefully place it in the top half of the test tube. Record what happens to the flame of the splint (match).
4. Balance the chemical reaction.
5. Identify the type of reaction.

Reactants	Appearance Prior to Reaction	Evidence of Reaction
HCl		
Na_2CO_3		

Observation of the burning match ______________________________

___Na_2CO_3 + ___ HCl → type:

Part F: Hydrogen Peroxide

Materials: Test tube, test tube rack, 3% H_2O_2, 0.1M KI (the potassium iodide is a catalyst in this reaction).

1. Place 3 mL of 3% H_2O_2 in a test tube. Record its appearance.
2. Place two dropperfuls of 0.1M KI in a second test tube. Pour the KI solution into the test tube that contains the H_2O_2 and record your observations.
3. Balance the chemical equation. Note KI is a catalyst, it only speeds up the reaction so is not involved in the balanced equation.

4. Identify the type of reaction.

Reactant	Appearance of Reactant	Evidence of a Reaction
H_2O_2		

____ $H_2O_2 \rightarrow$ type:

Questions and Problems

1. What is the evidence of a chemical reaction you might observe in the following?

a. Pouring vinegar into a dish that contains baking soda

b. Bleaching a stain

c. Burning a candle

d. Rusting of an iron nail

2. Balance the following equations:

a. ____ Na + ____ $Cl_2 \rightarrow$ ____ $NaCl$

b. ____ Mg + ____ $HCl \rightarrow$ ____ $MgCl_2$ + ____ H_2

c. ____ $Pb(NO_3)_2$ + ____ $KI \rightarrow$ ____ KNO_3 + ____ PbI_2

d. ____ $KClO_3 \rightarrow$ ____ KCl + ____ O_2

e. ____ C_3H_8 + ____ $O_2 \rightarrow$ ____ CO_2 + ____ H_2O

3. List the reaction type for each of the above equations:

a.

b.

c.

d.

e.

LAB 8

Moles and Chemical Formulas

We have now gained knowledge of the periodic table, electron configurations, and types of chemical reactions. In this lab, we will explore a different method of determining formulas for chemical compounds. when writing formulas for compounds, the simplest formula is preferred for ionic compounds. The simplest formula is one that has the lowest whole number ratio of all the elements contained in the compound. In the case of ionic compounds, the simplest formula is the only formula. With covalent molecules, the actual formula may be the simplest or it may be a whole number multiple of the simplest formula. For example, formaldehyde has a formula of CH_2O, the lowest whole number ratio of carbon to hydrogen to oxygen. Glucose, $C_6H_{12}O_6$, is a whole number multiple of this same formula. We determine the simplest ratio of a compound by converting the number of grams of each element into moles and then calculating the mole ratio of all the elements to each other. Let us calculate the simplest formula for a compound that contains zinc, Zn, and chlorine, Cl. In an experiment, it is determined that 2.616 g of Zn and 2.836 g of Cl combine to form zinc chloride. We begin by converting the mass of each element into moles by dividing by the molar mass of each (molar mass for an element is the same as that elements atomic mass).

$$\frac{2.616\ \text{g Zn}}{65.41\frac{\text{g}}{\text{mol}}\ \text{Zn}} = 0.040\ \text{mol Zn} \quad \text{and} \quad \frac{2.836\ \text{g Cl}}{35.45\frac{\text{g}}{\text{mol}}\ \text{Cl}} = 0.080\ \text{mol Cl}$$

Now that we have molar amounts of zinc and chlorine, we determine the lowest whole number ratio (simplest formula) by dividing each molar amount by the smallest number of moles, in this case 0.040 mol Zn.

$$\frac{0.040\ \text{mol Zn}}{0.040\ \text{mol}} = 1\ \text{Zn} \quad \text{and} \quad \frac{0.080\ \text{mol Cl}}{0.040\ \text{mol}} = 2\ \text{Cl}$$

The above calculations provide the ratio of one zinc for two chlorines and this becomes the subscripts in our formula for zinc chloride, $ZnCl_2$.

Along with simplest formulas for ionic and covalent compounds, we can also determine a formula for a hydrated salt. A hydrated salt, or hydrate, is a type of ionic compound that is hydrophilic, that is it absorbs water into its crystal lattice (much like a sponge absorbs water). Each hydrate has a fixed number of water molecules it can absorb. To show the number of water molecules a hydrated salt can hold, we show the formula of the salt and then follow with the number of water molecules it holds when hydrated. The dot linking the two formulas indicates loosely combined (not multiplication). See the following examples: $CaSO_4 \bullet 2H_2O$, read calcium sulfate dihydrate; $CuSO_4 \bullet 5H_2O$, read copper (II) sulfate pentahydrate; and $Na_2CO_3 \bullet 10H_2O$, read sodium carbonate decahydrate.

We determine experimentally the number of moles a hydrated salt can hold by dehydrating that salt. The water evaporates away when heated. We measure carefully the mass of the hydrated salt, then heat the substance until all the water is evaporated away; finally, we measure the mass of the anhydrous salt. As in the above example for calculating the simplest formula, we convert the mass of the anhydrate and mass of water into moles. Then we determine the mole ratio of water to the anhydrate. This should provide us with the coefficient of water in the formula.

Prelab Study Questions

1. What is meant by the simplest formula for a compound?

2. Differentiate between hydrate and anhydrate.

3. What process does one use to convert a hydrated salt into an anhydrous salt?

4. You measure the mass of a hydrate of $CoCl_2$ and record it as 6.00 g. After heating and then cooling the salt, the anhydrous form has a measured mass of 3.27 g.
   a. How many grams of water, H_2O, were lost in the heating (show your work)?

   b. Calculate the percentage (%) of water, H_2O, in the hydrate (show your work).

   c. Calculate the formula of the hydrate of $CoCl_2$ (show your work).

Part A: Determine the Simplest Formula

Materials: Crucible, crucible cover, hot mitt, clay triangle, iron ring clamp, ring stand, propane burner, magnesium ribbon, disposable pipette, small beaker, balance.

1. Clean and dry the crucible and cover. Once dried, heat the crucible and cover very gently for one minute to evaporate all the water that may be present. Let the crucible and cover cool completely before moving forward.
2. Place the crucible and cover on the balance and record the mass to the nearest 0.01 g.
3. Insure the magnesium ribbon at your station is clean of any tarnish. If there is tarnish, use steel wool to remove. Record the appearance of the magnesium ribbon.

4. Loosely coil the magnesium ribbon and place it in the bottom of the crucible. Place the crucible, cover, and magnesium ribbon on the balance and record the mass to the nearest 0.01 g.
5. Begin heating the crucible. Continue heating until you can determine that the magnesium is reacting with oxygen in the air (smoke or fumes and a bright light in the crucible indicate a reaction occurred). When the magnesium begins to burn, place the crucible cover (use tongs to do this) slightly offset on the crucible. When the magnesium stops producing smoke and flame, remove the cover and set it aside.
6. Continue strongly heating the crucible for another five minutes. Turn the propane burner off and allow the crucible to cool completely. During heating, some magnesium reacts with nitrogen in the air to form magnesium nitride. To remove the nitride product carefully add 15 to 20 drops of distilled water to the cooled crucible. This process produces ammonia, so do not inhale any of the fumes that are produced.
7. Cover the crucible again and heat gently for five minutes to drive off any excess water then heat strongly for five minutes longer. Allow the crucible to cool completely then place it on the balance and record the mass.

Part B: Formula of a Hydrate

Materials: Crucible, clay triangle, hot mitt, hydrate of magnesium sulfate, iron ring, ring stand, propane burner, balance.

1. Clean and dry the crucible and heat very gently for one minute to evaporate any water that may still be in it. Cool completely and place on a balance. Record the mass to the nearest 0.01 g.
2. Fill the crucible about one-third full of the hydrate of $MgSO_4$. Place the crucible with the hydrated salt and cover and record the mass to the nearest 0.01 g.
3. Place the crucible with the cover slightly offset on the clay triangle. Heat gently using the propane burner for 5 minutes, then increase the heat and continue heating for an additional 10 minutes. After 10 minutes, remove the burner and allow the crucible and contents to cool completely.

4. After cooling, place the crucible and cover on the balance and record the mass to the nearest 0.01 g. Repeat step 3 until 2 successive mass measurements are within 0.05 g.

Part A: Data Table

Mass of empty crucible and cover	g
Initial appearance of magnesium ribbon	g
Mass of crucible + cover + magnesium	g
Mass of crucible + cover + oxide product	g
Mass of magnesium (subtract out the crucible/cover	g
Mass of magnesium oxide product (again subtract out the crucible/cover)	g
Mass of oxygen in the compound (mass of product (- mass of magnesium ribbon)	g
Moles of magnesium (show calculation)	Mol
Moles of oxygen (show calculation)	Mol

Which element, Mg or O, has the smallest number of moles? ____________

Use this mole amount as the denominator in the calculations below that will determine the subscripts for the formula of the oxide product (show your work).

Formula based on your experimental calculations:

Part B: Data Table

1. Mass of crucible	g
2. Mass of crucible and hydrate	g

3. Mass of crucible and anhydrate (first heating)	g
4. Mass of crucible and anhydrate (second heating)	g
5. Mass of hydrate (step 2 – step1)	g
6. Mass of anhydrate (step 4– step 1)	g
7. Mass of water lost (step 5– step 6)	g
8. Percent of water in hydrate (show work)	%
9. Moles of water (show work)	mol
10. Moles of anhydrate (show work)	mol

Calculate the ratio of moles of water (step 9) to the moles of anhydrate salt (step 10), (show work)

Formula for the hydrate:

Write a balanced equation showing the dehydration of the above experimental formula for your salt.

LAB 9

Gas Laws

In lecture, we have discussed the relationship of pressure, temperature, volume, and the number of molecules of a gas. Several equations were developed to help explain these laws. Listed below are the three main equations used in relating these variables to the other.

Boyle's law relates how the change in pressure changes the volume of a gas.

$$P_1V_1 = P_2V_2$$

In this equation, P_1 and P_2 are the pressures of a gas and V_1 and V_2 are the volumes. Boyle's law states that as the pressure on a gas is increased, the volume must decrease.

Charles' law relates how a change in temperature results also in a change of volume of a gas.

$$\frac{V_1}{T_1} = \frac{V_2}{T_2}$$

In this equation, V_1 and V_2 are volume and T_1 and T_2 are temperature of a gas.

In both examples, the number of molecules of the gas must be kept constant. Further, we can combine both laws together for a combined gas equation.

$$\frac{P_1V_1}{T_1} = \frac{P_2V_2}{T_2}$$

When working with these three variables, pressure, temperature, and volume, we can use the combined gas equation to calculate what would happen when any one or two variables are altered.

We will investigate each of these gas law equations separately in this lab. The first is Boyle's law. Robert Boyle, keeping number of molecules of a gas constant,

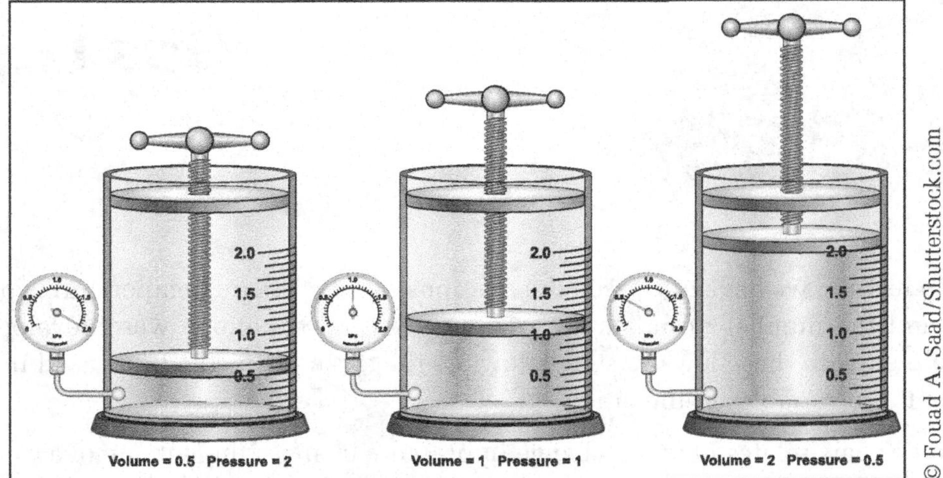

investigated what happens to the volume of that gas when the pressure of that gas is increased. Boyle discovered that the volume of a gas decreases proportional to the increase in the pressure. In other words, double the pressure on a gas and the volume will be half. By collecting numerous points of data, Boyle produced a graph of pressure as a function of volume and from this graph determined a constant relating the two variables. You will plot data of pressure and volume and determine Boyle's constant.

The second is Charles' law. Jacques Charles investigated how temperature would affect the volume of a gas. He determined that temperature was directly proportional to the volume of a gas. As the temperature is increased, the volume of the gas also increases. Charles used this relationship to graph temperature versus volume and, like Boyle, found that the slope of the line defined what is now known as Charles' constant. He was one of the early practioners of hot air ballooning.

Part A: Boyle's Law

Materials: Data and graph below.

1. In the data table below, calculate the $P \times V$ product for each coordinate set and record in the data table.
2. Plot the pressure (mmHg) of the gas in each reading against the volume (mL). The data will form a slight curve. Draw a smooth, best-fit line for the coordinate points on your graph.

Charles' Law

It is an experimental gas law that describes how gases tend to expand when heated, when the pressure is held constant.

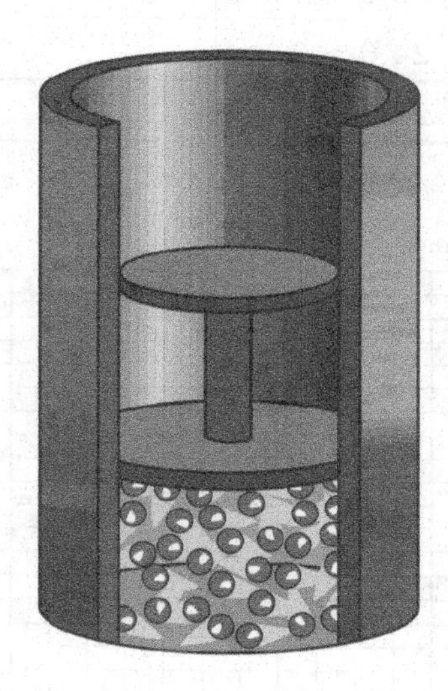

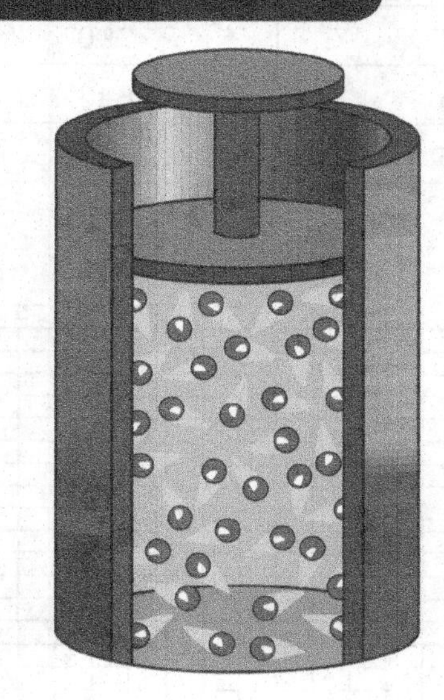

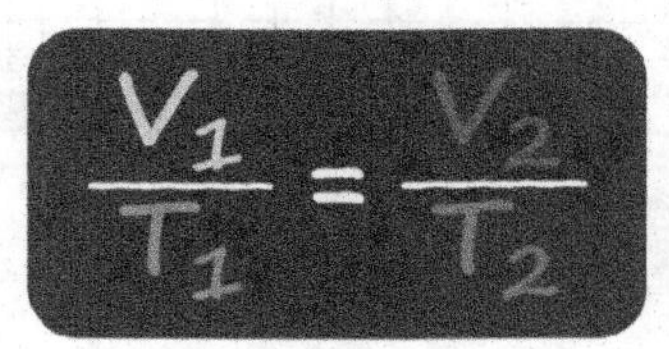

$$\frac{V_1}{T_1} = \frac{V_2}{T_2}$$ = *Constant, when the pressure is kept constant.*

Reading	Pressure (mmHg)	Volume (mL)	$P \times V$
1	630	32.0	
2	690	29.2	
3	726	27.8	
4	790	25.6	
5	843	24.0	
6	914	22.2	

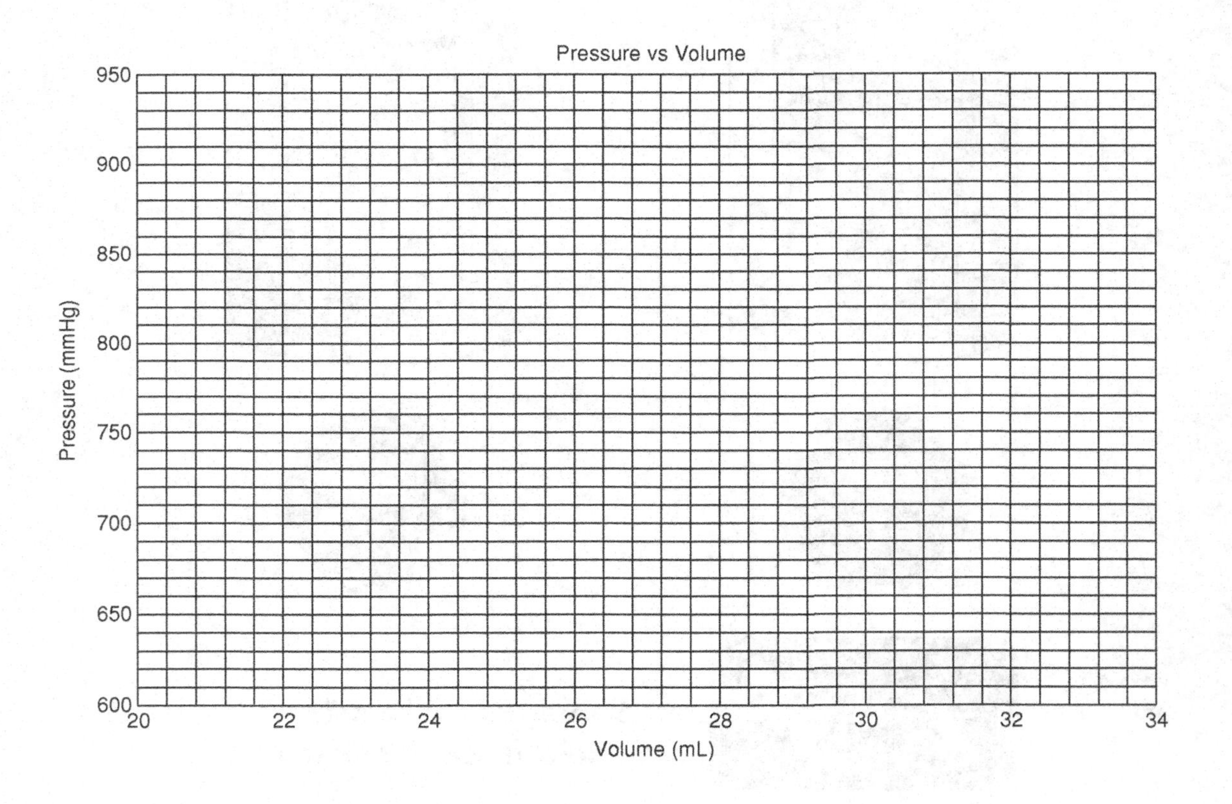

Part B: Charles' Law

Materials: 123-mL Erlenmeyer flask, 400-mL beaker, one-hole rubber stopper to fit the flask, water trough, thermometer, hot plate, ice, ring stand, and clamp for the flask.

1. Place the beaker on the hot plate and situate the ring stand so that it is next to the hot plate. Clamp the Erlenmeyer flask to the ring stand and lower it as far as the clamp will allow into the beaker. Fill the beaker with tap water so that most of the flask is covered, but so that water does not boil over.
2. Begin heating the water in the beaker. Boil gently for 10 minutes so that the air in the flask is about the same as the boiling water bath. Record the temperature of the boiling water bath just prior to removing the flask.
3. While the hot water bath is heating up the air in the flask, prepare a water trough with ice water.
4. Loosen the clamp from the ring stand, but do not remove it from the flask. Prior to removing the flask from the hot water bath, place your finger over the hole in the stopper. You must keep your finger over the hole until the flask is completely submerged in the cold-water bath.
5. Keeping your finger over the hole, place the flask with the clamp into the cold-water bath, completely submerging the flask. Once the flask is in the cold water, you may remove your finger. You should notice that some water is drawn up into the flask. Keep the flask submerged in the cold water for 10 minutes. This next part is crucial to equalize the pressure inside and outside of the flask.
6. Leaving the one-hole stopper open, you must adjust the flask below the water surface until the level of the water inside the flask is the same as the surface of the water in the water bath. Once the levels are equal, cover the hole in the stopper with your finger and keep it covered until you remove the flask and set it upright on the lab table.
7. Empty the contents of the flask into a graduated cylinder and measure the volume; record this volume in the data table. Now, fill the flask completely with water, place the stopper back in to squeeze out the extra water. Measure the total volume of water that the flask holds and record in the data table.

Temperature (°C)	
Temperature (K)	
Volume of cool water in flask	
Volume of water in flask	
Volume of air remaining in flask	
$\frac{V}{T}$	

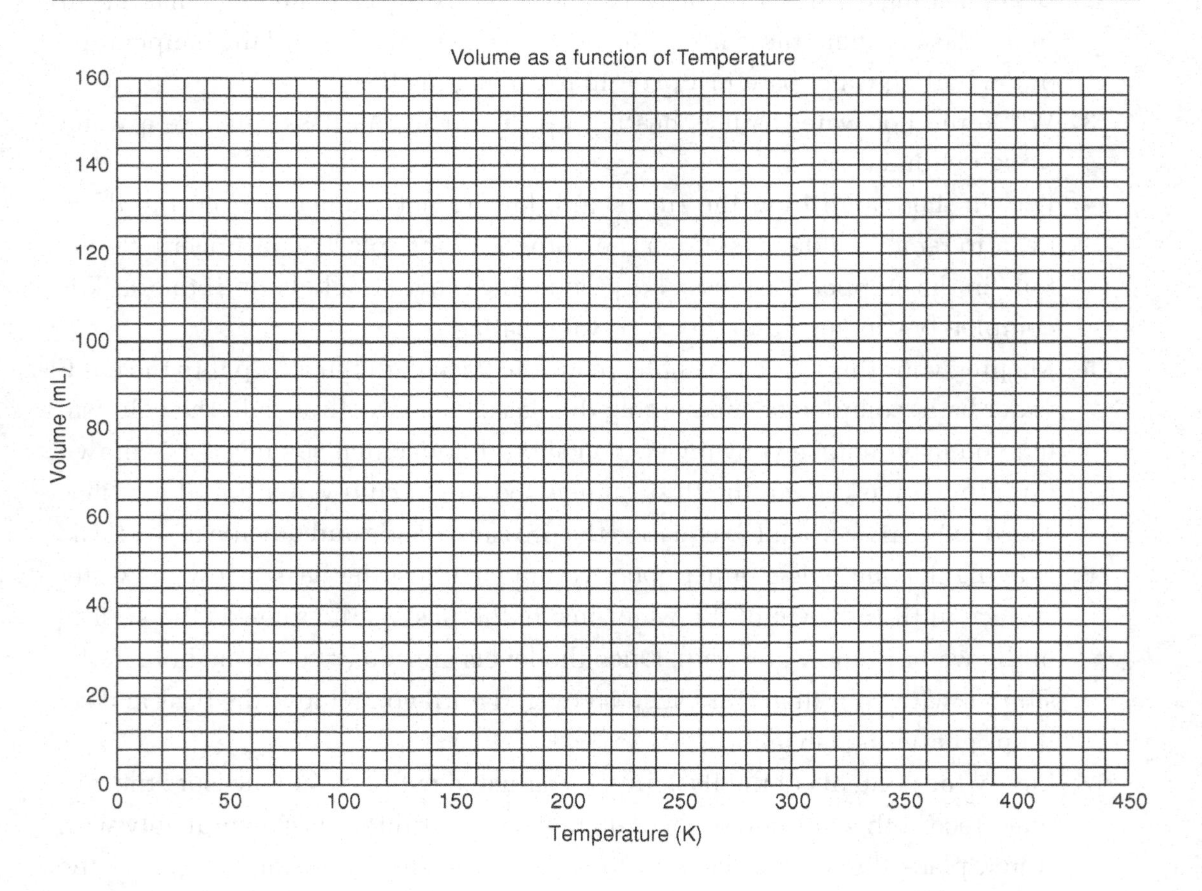

Questions and Problems

1. Look at your graph of pressure and volume. Is this a linear relationship? If yes, explain how it is linear, if not explain what type of relationship it is.

2. Using the pressure/volume graph, determine the volume of a gas at a pressure of 760 mmHg. Also, determine the pressure of a gas that has a volume of 30.0 mL.

3. A sample of helium gas has a volume of 234 mL at a pressure of 570 mmHg. What would the pressure be if you compress the volume to 150 mL? (show your work)

4. A sample of nitrogen has a volume of 75.0 mL at a pressure of 1.5 atm. What will be the volume of this same gas at a pressure of 0.75 atm? (show your work)

5. Using the graph of temperature/volume, what is the predicted Kelvin temperature of absolute zero? You will have to extend your line backward until it crosses the *x* axis.

6. How does your experimental value of absolute temperature compare to the accepted value of 0 K?

7. Contrast your temperature/volume graph to the graph of pressure/volume. What are the major differences? Which graph shows a linear relationship?

8. A gas with a volume of 525 mL at a temperature of 200 K is heated to 375 K. What is the new volume of the gas? (show your work)

9. A gas has a volume of 1.25 L at 298 K. At what temperature would the gas have a volume of 4.70 L? (show your work)

LAB 10

Solubility

If you drink coffee or tea, perhaps you do not really like the bitterness of the drink. So, you perhaps add a bit of sweetener to the hot beverage to make it a bit more palatable. You have just made a solution. The sugar, or other sweetener, dissolves into the coffee or tea. There are many different sorts of solutions out there today. For instance, soda pop is a solution of carbon dioxide dissolved into the liquid of the drink. The air we breathe is a solution of nitrogen, oxygen, and various other gases. If you like to go to the beach and swim in the ocean. . .yes, it too is a solution! In this lab, we will explore several different aspects of solutions especially how well something dissolves into solution or even if a substance can dissolve into solution.

When we make a solution, there are always two parts. There is the solvent (coffee or tea in the above example), that which does the dissolving. In addition, there is the solute (the sugar in the above example), that which is being dissolved. Usually, the solvent is the compound in greater abundance and the solute is the compound in lesser abundance. In order for these substances to combine to form a solution, both compounds must have similar polarity. A polar solvent, like water, can dissolve a polar solute, like salt or sugar. A nonpolar solvent, like gasoline, can dissolve a nonpolar substance, like Styrofoam. This is summed up in the adage "like dissolves like."

Whereas both molecules and ionic compounds can dissolve in water, ionic compounds go through a further process known as dissociation. In this process, the compound dissolves to the formula unit level and then the polar water molecules pull apart the formula unit into dissociated ions. The diagram to the left shows the dissociation of sodium chloride, NaCl, into sodium ions, Na^+, and chloride ions, Cl^-. The sodium cation is attracted to the polar negative end of a water molecule (the oxygen), while the chloride anion is attracted to the polar positive end of the water molecule (the hydrogen). This process is called hydration.

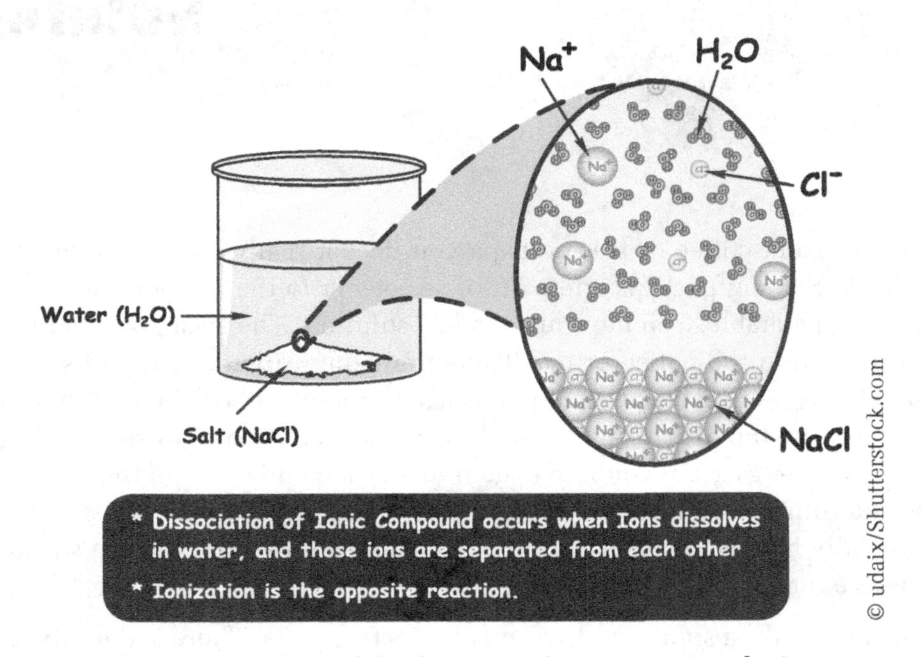

Because ionic compounds go through hydration to form an ionic solution, we can look at this solution in a special way when compared to molecules in solution. Ionic solutions are electrolytes and molecular solutions are nonelectrolytes. An electrolyte (as the name implies) is a good conductor of electricity and a nonelectrolyte solution is a poor conductor of electricity. The diagram shows a strong electrolyte that conducts electricity well lighting a bright lightbulb; a weak electrolyte conducts electricity somewhat with a dimly lit lightbulb; a nonelectrolyte does not conduct electricity, therefore the bulb remains unlit. NaCl is a strong electrolyte and sugar a nonelectrolyte.

Understanding electrolytic solutions in chemistry is important to understanding electrolytic solutions in our bodies. Electrolytes are important in cellular maintenance. Intracellular fluid has potassium as the major cation, K^+, and hydrogen carbonate, HCO_3^- as the major anion. Extracellular fluid contains sodium, Na^+, as the major cation and chloride, Cl^-, as the major anion. For osmotic pressure to work correctly in cellular maintenance, these electrolytic solutions must be balanced. Fluids added intravenously are isotonic, that is, the concentration of ions within the solution are the same as the ionic concentration in our bodies. Typically, the unit of measurement of concentration of these solutions is milliequivalents per liter (mEq/L).

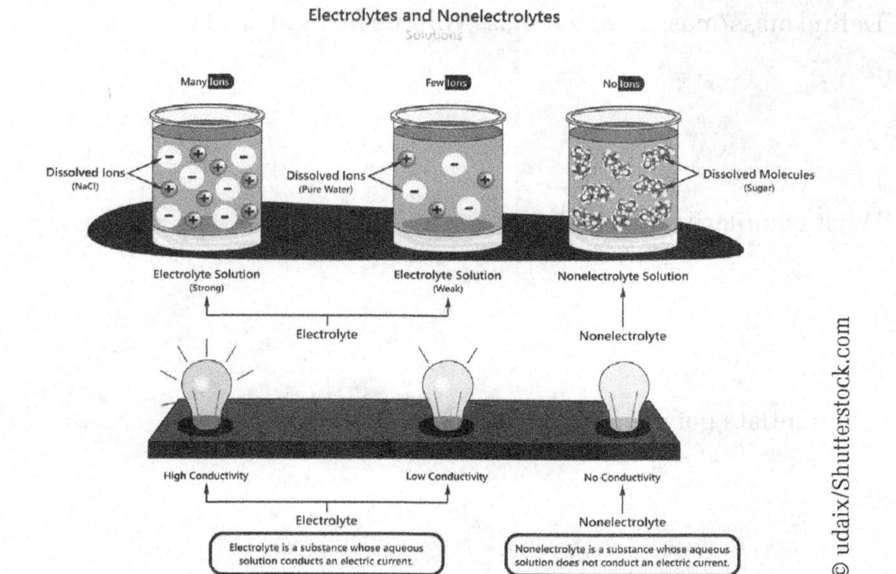

As mentioned above, we have units of measurement for solutions. mEq/L is one of these. We can also use molarity, moles solute per liter of solution, $\frac{\text{Mols}}{\text{L}}$, molality, moles of solute per kilogram of solution, $\frac{\text{Mols}}{\text{Kg}}$, or percentages. Percentage solutions can be expressed as percent by mass, percent mass to volume, or percent volume.

$$\text{Mass percent: } \frac{\text{Mass of solute (g)}}{\text{Mass of solutions (g)}} \times 100 = \text{Mass percent}$$

$$\text{Mass/volume percent: } \frac{\text{Mass of solute (g)}}{\text{Milliliters of solutions (mL)}} \times 100 = \text{mass / volume percent}$$

Prelab Study Questions

1. What are the consequences when water enters the gas tank of your car? (explain using the terms polar and nonpolar)

2. Define mass/mass percent, mass volume percent, and volume/volume percent.

3. What is molarity?

4. Differentiate between strong, weak, and nonelectrolytes.

5. You dissolve 3.26 g of KCl in enough water to make a 25.0-mL solution. Calculate the mass/volume percent for this solution (show your work).

Part A: Polarity of Solutes and Solvents

Materials: Test tubes, test tube rack, spatulas, stirring rods, KMnO4(s), I_2(s), sucrose (s), vegetable oil, cyclohexane.

Caution: iodine, I_2, can **burn the skin**, handle carefully! Cyclohexane is *flammable*, do not handle around open flames!

1. Place the test tubes in a test tube rack. To 4 test tubes, add 3 mL of distilled water (polar solvent). To the remaining 4 test tubes, add 3 mL of cyclohexane (nonpolar solvent). Add a small amount $KMnO_4$ into a test tube that contains water and into a test tube that contains cyclohexane. Stir with the stirring rod and note your observations in the data table. Repeat this procedure with each

of the other solutes, I_2, sucrose, and vegetable oil. Make sure you clean and dry the stirring rod between each test tube.

2. Based on your observations identify each solute as polar or nonpolar.

Part B: Electrolytes and Nonelectrolytes (Teacher Demonstration)

Materials: Conductivity apparatus, small beakers, 0.1 M NaCl, 0.1 M sucrose, 0.1 M HCl, 0.1 M $HC_2H_3O_2$ (acetic acid), 0.1 M NaOH, 0.1 M NH_4OH (ammonium hydroxide), 0.1 M C_2H_5OH (ethanol).

1. Place 15 to 20 mL of the above solutions into small beakers. Carefully lower the conductivity apparatus into each solution and note whether there is a tone or not, and if there is a tone, note how loud the tone is in the data table. A tone indicates an electrolytic solution, a loud tone indicates a strong electrolytic solution, and no tone indicates a nonelectrolytic solution.
2. Indicate in the data table whether each solution is a strong, weak, or nonelectrolyte.
3. Indicate the types of particles present in each solution.

Part C: Electrolytes in Body Fluids

Materials: Samples of intravenous (IV) solutions such as saline, lactated ringer's solutions, sucrose, and so on.

1. Select several of the IV solution samples in the laboratory and record the type of solution in each
2. List the cations and their concentrations in mEq/L
3. List the anions and their concentrations in mEq/L
4. Calculate the total number of mEq of cations
5. Calculate the total number of mEq of anions
6. Calculate the overall sum of cations and anions in each IV solution

Part D: Concentration of a Sodium Chloride, NaCl, Solution

Materials: Hot plate, evaporating dish, 400-mL beaker, 10-mL graduated cylinder, NaCl solution.

Source: Michael O'Donnell

1. Place the evaporating dish on the balance and record the mass in the data table.
2. Measure out 10.0 mL sample of the NaCl solution and record the volume in the data table. Next pour the solution into the evaporating dish, place it on the balance, and record the total mass of the dish and solution in the data table.
3. Fill the beaker about half full with tap water and place the beaker on the hot plate. Place the evaporating dish on top of the beaker so that the spout of the beaker and evaporating dish are aligned. Turn on the hot plate and heat until boiling (you may need to carefully add more water to the beaker). The solution in the evaporating dish should begin to evaporate away. When the contents of the evaporating dish appear dry, turn off the hot plate and remove the beaker and evaporating dish. When the evaporating dish has cooled, place it alone back on the hot plate. Turn the hot plate to a low temperature and continue heating the evaporating dish until the NaCl is completely dried. Remove the evaporating dish and allow to completely cool.
4. Place the evaporating dish on the balance and record the final mass.

Part A: Polarity of Solutes and Solutions

Solute	Soluble/Insoluble in:		Polar or Nonpolar Solute
	Water	Cyclohexane	
$KMnO_4$			
I_2			

Solute	Soluble/Insoluble in:		Polar or Nonpolar Solute
	Water	Cyclohexane	
Sucrose			
Vegetable oil			

Part B: Electrolyte and Nonelectrolyte

Solution	Intensity of Sound (No Sound, Weak Sound, Loud Sound)	Type of Electrolyte (Nonelectrolyte, Weak, Strong)	Type of Particle (Ions, Molecules, Both)
0.1 M NaCl			
0.1 M Sucrose			
0.1 M HCl			
0.1 M $HC_2H_3O_2$ (acetic acid)			
0.1 M NaOH			
0.1 M NH_4OH			
0.1 M C_2H_5OH (ethanol)			

Part C: Electrolytes in Body Fluids

Type of intravenous (IV) solution			
Cations (mEq/L) List each cation and amount			
Anions (mEq/L) List each anion and amount			
Total charge of cations			
Total charge of anions			
Sum of cations and anions			

Part D: Concentration of Sodium Chloride Solution

Mass of evaporating dish	g
Volume of NaCl solution	mL
Mass of dish and NaCl solution	g
Mass of dish and dry NaCl	g

Calculations	
Mass of NaCl solution	g
Mass of NaCl dry salt	g
Mass/mass percent (show calculations)	
Mass/volume percent (show calculations)	
Moles of NaCl (show calculations)	mol
Volume of solution in liters	L
Molarity of NaCl solution (show calculation)	mol/L

Questions and Problems

1. Why are some solutes only in water, while other are soluble only in cyclohexane?

2. What should be the overall charge of any intravenous (IV) solution? Why?

3. A 15.0-mL sample of NaCl solution has a mass of 15.78 g. After the solution is evaporated to dryness, the dry salt residue has a mass of 3.26 g. Calculate the following concentrations for this solution:
   a. Mass/mass percent

   b. Mass/volume percent

   c. Molarity (M)

4. How many grams of KI are in 25.0 mL of a 3.0% m/v KI solution? Show your calculations.

LAB II

Cations and Anions: How Do We Know What's Present?

As we discussed in the previous lab, when we make a solution of an ionic compound, not only does the compound dissolve, but it also dissociates into ions. We further defined a cation as a positively charged particle and an anion as a negatively charged particle. However, short of telling you what ions are present in a compound, how do we know? This lab will present several methods we use to test for the presence of unknown ions contained in a solution. We will test several known ionic solutions as well as an unknown solution. In the end, you should be able to identify the ions contained in the unknown solution.

The first of these tests is the flame test. Ions, when held in a burner flame, will emit a particular color of flame unique to that ion. This color depends on the metal element that is present as a cation. As the compound is held in a burner flame, the metal ion absorbs energy from the flame and will then emit a color as the electrons that were excited release the absorbed energy in the form of light. The metals we tested in this lab, Na^+, K^+, and Ca^{2+}, emitted a distinctive color during the flame test.

Another test that we use will help detect the presence of the ammonium ion, NH^{4+}, when it converts to ammonia, NH_3. In this test, a distinct odor is emitted and red litmus paper turns blue. A final test for cations determines the presence of the cation for iron, Fe^{3+}. When a solution with the Fe^{3+} ion is added to a solution containing potassium thiocyanate, KSCN, a distinctive red color forms.

Just as there are tests for cations, there are tests for anions. Silver nitrate, $AgNO_3$, is often used to detect the presence of several anions. For instance, when a silver nitrate solution is added to a solution containing a chloride, such as sodium chloride, a precipitate forms in the reaction. AgCl is insoluble. Adding nitric acid, HNO_3, can confirm the presence of chloride since the chloride compound will not dissolve. Together, this gives a positive test for the presence of silver chloride.

There is a positive test for the phosphate ion, PO_4^{3-}, as well. Ammonium molybdate and any phosphate compound will form a yellow precipitate. We use barium chloride, BaCl, as a positive test for the sulfate anion, SO_4^{2-}. The precipitate that forms is barium sulfate. Barium can form precipitate salts with other anions, but adding HNO_3 will cause these precipitates to dissolve. Finally, adding hydrochloric acid, HCl, to any solution that contains the carbonate anion, CO_3^{2-}, forms gas bubbles of CO_2.

Once all of these tests are completed, you will determine the formula of your unknown substance. We use these tests and others for any unknown out there as a way to determine the cations and anions present. This is not an exhaustive means to determine ions in compounds but covers many common ions. As a final part of this lab, you may be asked to work with some common household items (cleaners, sodas, etc.) and determine whether any of these common ions are present.

Procedures

Part A: Flame Tests for Cations

Materials: Wooden splints soaked in saturated solutions of $CaCl_2$, KCl, NaCl; a soaked wooden splint of an unknown; propane burner. CAUTION, using a propane burner and wooden splints is a fire hazard, pay close attention!

1. Carefully hold the end of each of the known ion-soaked wooden splints in the blue flame of the propane burner and record the color observed.
2. Hold the soaked unknown wooden splint in the propane flame and record the flame color observed. If the color matches any of the three known solutions, you may record that as your unknown cation.

Part B: Tests for NH^{4+} and Fe^{3+} Cations

Materials: Test tubes, test tube rack, 6 M NaOH, red litmus paper, 6 M HNO_3, 0.1 M KSCN, 0.1 M NH_4Cl, 0.1 M $FeCl_3$, warm water bath, stirring rod.

1. **Test for NH^{4+}**: Place two droppers full of NH_4Cl in a test tube and two droppers full of unknown in a second test tube. Add 15 drops of NaOH to each test tube and carefully waft the vapors toward you. You should notice a smell of

ammonia if the NH^{4+} cation is present. Place a strip of moistened red litmus paper across the top of each test tube and place the test tubes in a beaker of warm water. Do not let the litmus paper touch the NaOH. If any ammonia is given off, the litmus paper will turn blue. Record your observations of the known and unknown.

2. **Test for Fe^{3+}**: Place two droppers full of 0.1 M $FeCl_3$ in a test tube and two droppers full of your unknown in another test tube. Add 5 drops of 6 M HNO_3 to each test tube. Then, add 2 to 3 drops of KSCN to each test tube. If the Fe^{3+} cation is present, the solution will turn a deep red color. Record your observations.

Part C: Tests for Anions

Materials: Test tubes, test tube rack, 0.1 M $AgNO_3$, 3 M HCl, 6 M HNO_3, stirring rod, 0.1 M Na_2SO_4, 0.1 M $BaCl_2$, 0.1 M Na_3PO_4, $(NH_4)_2MoO_4$ (ammonium molybdate reagent), 0.1 M Na_2CO_3, hot water bath.

1. **Test for Chloride, Cl^-, anion**: Place two droppers full of 0.1 M NaCl solution in a test tube and two droppers full of your unknown in another test tube. Add 5 to 10 drops of 0.1 M $AgNO_3$ to each test tube. Add 10 drops of 6 M HNO_3 to each test tube. Stir with the stirring rod. Any white precipitate that remains contains the Cl^- anion. Record your observations.
2. **Test for sulfate, SO_4^{2-}, anion**: Place two droppers full of 0.1 Na_2SO_4 solution in a test tube and two droppers full of your unknown in another test tube. Add one dropper full of $BaCl_2$ to each test tube. Add 10 drops of 6 M HNO_3 to each test tube. Any white solid that remains contains the SO_4^{2-} anion.
3. **Test for phosphate, PO_4^{3-}, anion**: Place two droppers full of 0.1 M Na_3PO_4 solutions in a test tube and two droppers full of your unknown solution in another test tube. Add 10 drops of 6 M HNO_3 to each test tube. Warm the test tubes in a hot water bath (60°C). To each test tube add 15 drops of ammonium molybdate solution, $(NH_4)_2MoO_4$. The formation of a yellow precipitate indicates the presence of the PO_4^{3-} anion. Record your observations.

Part D: Household Product Testing

Materials: Any household product your instructor has provided.
To conduct a test of household products, prepare a solution of what is available. Then, use the above tests to determine what cations and anions are present in your sample.

Prelab Study Questions

1. Describe some of the methods that can be used to detect the presence of ions in solution.

2. What tests could you use to identify a solution of Ag_3PO_4, silver phosphate?

3. What tests could you use to identify a solution of $FeCl_3$, iron (III) chloride?

4. You have an unknown colorless solution that provides a yellow flame during a flame test. You conduct a further test using $AgNO_3$ and observe a white precipitate form that remains after you add HNO_3. Identify the two ions that are present and name the compound.

Part A: Flame Tests

Cation Tested	Color Produced
K^+	
Ca^{2+}	
Na^+	
Unknown	

Identification of unknown: Based on the above tests does your unknown contain the above ions, K^+, Ca^{2+}, Na^+, or none of the above? Explain your answer.

Part B: Tests for NH^{4+} and Fe^{3+}

Cation	Test Results	
	Odor	Litmus
NH^{4+}		
Fe^{3+}		
Unknown		

Identification of unknown: Based on the above tests does your unknown contain either NH^{4+} or Fe^{3+}, or none of these? Explain your answer.

Part C: Tests for Anions

Anion	Observations of Known Solution	Observations of Unknown Solution
Cl^-		
SO_4^{2-}		
PO_4^{3-}		
CO_3^{2-}		

From the observation of anion tests above, which anion does your unknown contain. Explain your answer.

Part D: Writing the Unknown Salt Formula

Cation present__________ Name__________

Anion present__________ Name__________

Formula for the unknown salt__________________

Name for the unknown salt__________________

Part E: (If Required by the Instructor) Testing Consumer Products for Cations and Anions

Cation Tests	Observations	Ion(s) Present
Flame tests (K^+, Ca^{2+}, Na^+)		
NH^{4+}		
Fe^{3+}		
Anion tests		
Cl^-		
SO_4^{2-}		
PO_4^{3-}		
CO_3^{2-}		

Questions and Problems

1. Why do you test known cation and anion solutions in order to test an unknown solution to identify the compound?

2. You have an unknown solution that contains either NaCl or $CaCl_2$. What tests would you run to identify the compound?

3. If a solution turns red when a few drops of KSCN are added, what cation is present?

4. Looking at a label for plant food you see that it contains the compound $(NH_4)_3PO_4$. What tests would you run to verify the presence of the NH^{4+} cation and the PO_4^{3-} anion?

LAB 12

Reaction Rates and Chemical Equilibrium

We have observed several different reaction types through the semester. In most cases, these reactions happened quickly, the exception being an endothermic decomposition. This particular reaction can take some time, as the heat supplied with a propane burner is necessary throughout the entire reaction. There are methods to speed up the rate at which a reaction proceeds. A chemical reaction occurs as molecules and ions collide; these collisions break apart the original substances and form new bonds. If we create a better chance of these collisions occurring, we speed up the reaction. Three factors cause a reaction to speed up. The first is temperature. By increasing the temperature of a reaction, the speed of molecules and ions increases leading to more collisions. An example of this takes place in kitchens around the world. We heat food to make it cook faster. Our body temperature increases when we are sick, speeding up the metabolic reactions. The second factor is concentration of reactants. If more of a reactant is present, the collisions between molecules and ions occurs more often. If we remove some reactant, we slow down the reaction rate. Finally, we can use catalysts to speed up a reaction. A catalyst is an element or compound that lowers the activation energy of a reaction allowing it to speed up. Enzymes are biological catalysts that make cellular reactions occur at rates needed for cellular survival. Industry uses catalysts to speed up reactions that produce needed compounds. In all of the above instances, the rate of the reaction speeds up so that we get the products in a timely manner.

These reactions typically all proceed in one direction, which is from reactants to products. In this lab, we also will observe reactions that proceed from reactants to products only until the concentration of products reaches a certain point. At that time, the reaction begins to reverse. These are termed equilibrium reactions. Once the reaction achieves equilibrium, the reaction proceeds both forward toward products and reverse toward reactants. We represent equilibrium reactions in a slightly different manner than other reaction types. We use a double arrow that shows the reaction proceeding forward and backward as in the below example.

$$2SO_2 + O_2 \leftrightarrow 2SO_3$$

At equilibrium, we can calculate an equilibrium constant according to Le Châtelier's principle. The equilibrium constant is based on the concentrations of the reactants and products. Any coefficient in the balanced equation is an exponent for the concentration. The equilibrium constant indicates the extent of the forward reaction. A K_c greater than 1 indicates a reaction that tends to the product side of the reaction. A K_c value less than 1 indicates a reaction that tends to the reactant side of the reaction.

$$K_c = \frac{[\text{Products}]}{[\text{Reactants}]}$$

Alternatively, using the above reaction as an example:

$$K_c = \frac{[SO_3]^2}{[SO_2]^2[O_2]}$$

Once at equilibrium, we begin to change those conditions in several ways. Since there is a balance of concentrations of reactants and products, equilibrium changes when we alter one of these concentrations. Changing these concentrations shifts the reaction either toward the product side or toward the reactant side. Le Châtelier's principle allows us to predict how the reaction shifts.

$$A_a + B_b \leftrightarrow C_c + D_d$$

Using the above general reaction as an example, the following equilibrium shifts are possible according to Le Châtelier's principle:

An increase in the concentration of *A* or *B* causes a shift in equilibrium to the right, or the products.
A decrease in the concentration of *C* or *D* causes a shift in equilibrium to the right, or the products.
An increase in the concentration of *C* or *D* causes a shift in equilibrium to the left, or the reactants.
A decrease in the concentration of *A* or *B* causes a shift in equilibrium to the left, or the reactants.

It is through observations that we determine whether an equilibrium change occurs. Typically, equilibrium reactions involve a change of color.

Prelab Study Questions

1. What factors increase the rate of a chemical reaction?

2. You have a system in equilibrium. What would happen to the equilibrium if you add more reactant to the system?

3. You have a system in equilibrium. What would happen to the equilibrium if you add more product?

4. Write the equilibrium constant equation for the following reaction:
   a. $H_2 + I_2 \leftrightarrow 2\ HI$

   b. $2\ NO_2(g) \leftrightarrow N_2O_4(g)$

5. If you determine an equilibrium constant of $K_c = 10$ for a reaction, what does this indicate about the reaction?

Part A: Rate of Chemical Reactions

A.1 Effect of Temperature on Reaction Rate

Materials: Test tubes, beaker (400 mL preferred), ice, 0.1 M HCl, $NaHCO_3$, thermometer, hot plate.

1. Pour 10 mL of 0.1 M HCl into 2 test tubes. Place 1 test tube in a beaker of warm water (50°C–60°C) and place the other test tube in a beaker of ice water (10°C or less).
2. Measure the temperature of each sample and record. Place a small amount (the end of a scoopula) of $NaHCO_3$ in each test tube.
3. Observe the fizzing in each test tube. Record which test tube clears first (indicating the end of the reaction).

A.2 Effect of Concentration on Reaction Rate

Materials: Test tubes, 3 clean pieces of magnesium ribbon (2–3 cm long), 1.0 M HCl, 2.0 M HCl, 3.0 M HCl, timing device.

1. Label each of three test tubes so that you can track what is in each. Place one piece of magnesium into each of the test tubes.
2. Add 10.0 mL of 1.0 M HCl into the first test tube and record the time it takes for all of the magnesium ribbon to be consumed.
3. Repeat the above test with the second test tube and 2.0 M HCl and then in the last test tube with 3.0 M HCl.

A.3 Effect of a Catalyst

Materials: Test tubes, 3% H_2O_2 (peroxide), MnO_2, piece of Zn, freshly cut potato, piece of boiled potato.

1. Place five test tubes in a rack and mark them in a way to track each part of the experiment.
2. Pour 2 mL of 3% peroxide into each test tube.
   a. Test tube 1 is the reference
   b. Test tube 2 add a small amount (scoopula end) of MnO_2

c. Test tube 3 add a small piece of Zn
d. Test tube 4 add a small piece of fresh potato
e. Test tube 5 add a small piece of boiled potato
3. Record your observations of each test tube

Part B: Reversible Reactions

B.1 Hydrates

Materials: Test tube, $CuSO_4{\cdot}5H_2O$ (copper sulfate pentahydrate), propane burner, test tube holder, small beaker with distilled water.

1. Record the initial appearance of the copper sulfate. Using the test tube holder, and ensuring that the open end of the test tube is not pointing at anyone, heat the test tube over a low flame. Move the test tube constantly. When some of the sample has turned from blue to white, observe the part of the test tube closest to the open end for drops of moisture.
2. Turn off the burner and allow the test tube to cool to room temperature. Once cool, use a dropper to add water to the white powder in the test tube. Record your observations.

B.2 Reaction of Copper (II) Ion and Hydroxide

Materials: Test tubes, rack, droppers, stirring rod, 0.1 M $CuCl_2$, 0.1 M NaOH, 0.1 M HCl.

1. Place 2 droppers full of 0.1 M $CuCl_2$ in each test tube and label as 1 and 2. Record the color of the solution.
2. Add drops of 0.1 M NaOH to test tube 1 and stir until a light blue precipitate of $Cu(OH)_2$ first forms. Add the same number of drops to test tube 2. Record your description of the initial appearance of the contents of the test tubes.
3. In test tube 1 continue to add drops of 0.1 M NaOH. This increases the available OH^-. Describe any change to test tube 1.
4. Add drops of 0.1 M HCl to test tube 2, which decreases availability of OH^-. Describe any change in the appearance in test tube 2.

Part C: Le Châtelier's Principle

Materials: Test tubes, rack, small beaker, 0.01 M $Fe(NO_3)_3$, 0.01 M KSCN, 1 M $Fe(NO_3)_3$, 1 M KSCN, 3 M HCl, beaker for warm water bath, beaker for cold water bath, ice, hot plate.

In this part of the experiment, we will stress an equilibrium system that contains Fe^{3+}, SCN^-, and $FeSCN^{2+}$ ions. As the concentration of each ion changes, the direction of equilibrium will shift. After changing concentrations, we will then look at how temperature stresses this same system. The following is the equilibrium equation:

$$Fe^{3+}(aq) + SCN^-(aq) \leftrightarrow FeSCN^{2+}(aq)$$

Yellow colorless red

C.1 Effect of Concentration

1. Measure 10.0 mL of 0.01 M $Fe(NO_3)_3$ and 10.0 mL of KSCN into a small beaker and stir.
2. From this stock solution, add 3 mL to each of 6 test tubes. Then, follow the below instructions for the individual test tubes.
   a. Test tube 1: add 10 drops of distilled water, stir and record the color observed. This is a reference test tube.
   b. Test tube 2: add 10 drops of 1 M $Fe(NO_3)_3$, stir and record the color you observe compared to the reference.
   c. Test tube 3: add 10 drops of 1 M KSCN, stir and record the color you observe compare to the reference.
   d. Test tube 4: add 10 drops of 3 M HCl, stir and record the color you observe compared to the reference

C.2 Effect of Temperature

1. Place test tube 5 from above in the ice water bath. Cool to a temperature below 10°C. Add 10 drops of distilled water and stir.
2. Place test tube 6 from above in the warm water bath on the hot plate. Heat to between 50°C and 60°C. Turn off the hot plate and add 10 drops of distilled water.

3. After 10 minutes, remove the test tubes from the cold and hot water baths and record your observations of color compared to the reference above.
4. State whether the $FeSCN^{2+}$ increased or decreased in test tube 5 and 6.

Part A: Rate of Reaction

A.1 Temperature

Test Tube	Temperature	Observation
1		
2		
Which cleared first:	1	2

A.2 Concentration

Test Tube	Concentration of HCl	Initial Time	Final Time	Total Time
1	1.0 M			
2	2.0 M			
3	3.0 M			

A.3 Catalyst

Test Tube	Observation	Catalyst Present	
Reference		Yes	No
MnO_2		Yes	No
Zn		Yes	No
Fresh potato		Yes	No
Boiled potato		Yes	No

Part B: Equilibrium

B.1 Hydrate

	Observations
Initial hydrate appearance	
Hydrate appearance after heating	
Hydrate appearance after adding water	

B.2 Cu^{2+} and Hydroxide

	Test Tube 1	Test Tube 2
Initial color $CuCl_2$ solution		
Initial appearance of $Cu(OH)_2$ precipitate		
Change in appearance		
Change in appearance		

Part C: Le Châtelier's Principle

C.1 Concentration

Test tube	Color	Deeper or Lighter Red (Compared to Reference)		$FeSCN^{2+}$ Increases or Decreases	
Reference					
Fe^{3+} added		Lighter	Darker	Increase	Decrease
SCN^- added		Lighter	Darker	Increase	Decrease
Fe^{3+} removed		Lighter	Darker	Increase	decrease

C.2 Temperature

Test tube	Color	Temperature (Increase or Decrease)		Color Deeper or Lighter than Reference		$FeSCN^{2+}$ Increase or Decrease	
Reference							
Ice water		Increase	Decrease	Lighter	Darker	Increase	Decrease
Warm water		Increase	Decrease	Lighter	Darker	Increase	Decrease

Questions and Problems

1. How did an increase in temperature effect the rate of reaction?

2. How did an increase in the concentration of HCl cause a difference in the rates of reaction with Mg?

3. Which catalyst in **Part A.3** was most effective?

4. How can you explain the difference in catalytic activity of the fresh versus boiled potatoe?

5. From where did the water droplets you observed at the mouth of the test tube in **Part B.1** come?

6. When the copper (II) sulfate pentahydrate was heated, did the reaction shift toward the product side of the equilibrium or the reactant side? Explain.

7. Describe what happened when you added water to the residue from heating the copper (II) sulfate pentahydrate and indicate whether the equilibrium shifted toward the product or reactant side.

8. In which direction did the reaction shift when Fe^{3+} ions were added to the system? What evidence supports this decision?

9. List the evidence that supports the equilibrium shifted in the direction of the products when temperature was decreased in **Part C.2**.

LAB 13

Acids and Bases and Buffers

What is happening when you have heartburn? Why do you take Tums or Rolaids to relieve indigestion? What is meant by buffered aspirin? What is meant by the term "acid rain"? How do we deal with acid mine drainage (amd)? This lab will look at these issues by looking at acids and bases.

There are several ways to define an acid or a base. The first of these definitions is from Arrhenius. An Arrhenius acid is a solution that has an excess of H^+ and an Arrhenius base is a solution that has an excess of OH^-. These ions are from the substance in solution. Ionic compounds dissociate after dissolving into solution. When you bubble HCl(g) through water, some of the HCl dissolves and dissociates into H^+ and Cl^-. This is an Arrhenius acid. Or, if you dissolve sodium hydroxide, NaOH, into solution, it dissociates into Na^+ and OH^-, which is an Arrhenius base. Hydrochloric acid, HCl, is a strong acid used in many laboratory reactions. Sodium hydroxide, NaOH, is a very common strong base found not only in the lab, but also in household chemicals such as drain cleaners.

We can measure the strength of an acid or base by measuring the pH of the solution. The lower the pH, the more acidic the solution. pH is the negative log of the concentration of the hydronium ion, $[H_3O^+]$. Though there are instruments to measure pH directly (a pH meter and red cabbage indicator are used in this lab), we can determine the $[H_3O^+]$ and then use the following formula to calculate the pH:

$$pH = -\log[H_3O^+]$$

With pure water, the concentration of H_3O^+ and OH^- equal 1.0×10^{-7}, or

$$[H_3O^+] = [OH^-] = 1.0 \times 10^{-7}$$

Given this relationship, we can calculate $[H_3O^+]$ if we know $[OH^-]$ by rearranging the equation to solve for $[H_3O^+]$. The pH of a solution ranges from 0, very acidic, to 14, very basic.

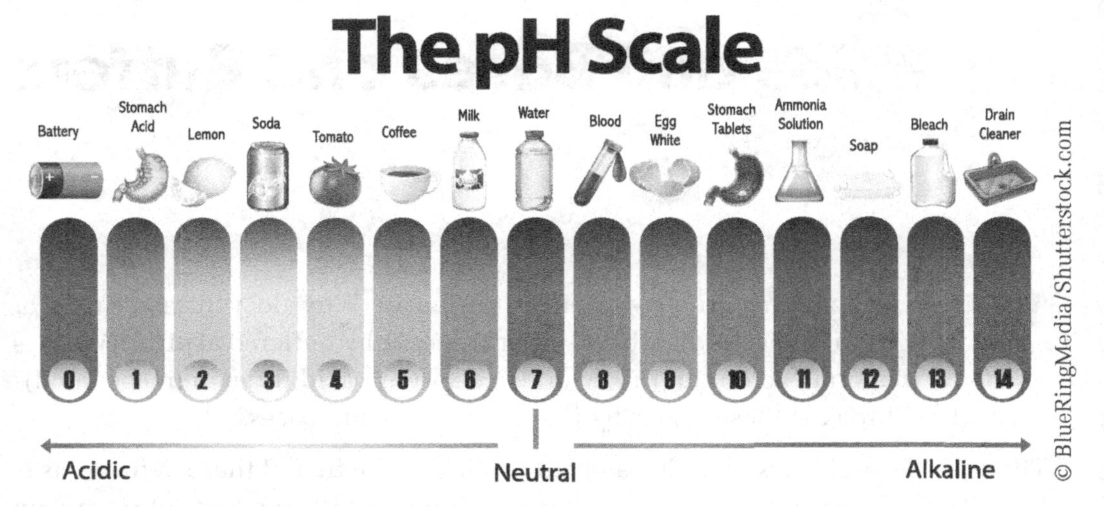

Now, what about that idea of buffered aspirin? There are certain solutions that we call buffered. These solutions resist large changes in pH if either an acid or base is added to the solution. Buffered aspirin does not allow the pH of the stomach to increase from the acetylsalicylic acid, which is the active ingredient in aspirin. Our blood pH is maintained between 7.35 and 7.45 by buffers in our body. A drop in the blood pH leads to acidotic blood, which can cause damage to the cells. A buffer maintains a certain pH of a solution by neutralizing small amounts of either acids or bases that are added. Many buffers contain a weak acid with its salt. The weak acid neutralizes any added base, and the anion of the salt reacts with excess H^+ to neutralize any additional acid.

Prelab Study Questions

1. What is the equation used to determine the pH of a solution?

2. In the following solutions, state whether the pH indicates a basic, acidic, or neutral solution.

   a. pH = 2.5

   b. pH = 7.3

c. pH = 9.7

d. pH = 7.0

3. If the hydronium ion, $[H_3O^+]$, has a concentration of 0.0032 M, what is the pH of that solution? (show your work). Is the solution acidic, basic, or neutral?

4. You determine the concentration of $[OH^-]$ as 0.05 M. Calculate $[H_3O^+]$. Then calculate the pH of this solution. Is the solution acidic, basic, or neutral? (show your work)

Part A: Measuring pH

Materials: Test tubes, samples to test (provided by your instructor), cabbage juice indicator, and pH reference standards (provided by instructor), pH meter, calibration buffers, wash bottle.

1. Place a dropper full of each sample to test into separate test tubes. If the sample is solid or viscous, mix with distilled water to form a solution. Add a dropper full of red cabbage indicator to each test tube. Shake the test tube to mix, then record the color of the sample.
2. Compare the color of the sample to the reference set of indicators and record the pH of the sample.
3. Using a small beaker into which the pH probe can fit, add about 2 mL of each sample solution to the beaker. Using the pH meter, measure and record the pH of each sample. Rinse off the pH probe with distilled water between each sample.

Part B: Effect of Buffers

Materials: High pH buffer (9–11), low pH buffer (3–4), droppers, test tubes, 0.1 M NaCl, 0.1 M HCl, 0.1 M NaOH, cabbage juice indicator.

Set up 2 sets of 4 test tubes with the following: (use a set for steps 1 through 5 below, and the second set for steps 6 through 10).

10 mL of H_2O, 10 mL of high pH buffer, 10 mL of 0.1 NaCl, 10 mL of low pH buffer

1. Using a dropper full cabbage indicator, determine the pH of each of the solutions in the four test tubes and record.
2. Add 5 drops of 0.1 M HCl to each of the first set of test tubes and mix. Determine the pH of each solution using the cabbage indicator and reference set.
3. Add 5 more drops of 0.1 M HCl to each test tube and record the new pH of each.
4. Determine the change in pH for each solution.
5. Identify which solutions are buffered.
6. Using the second set of four test tubes, record the pH of each using the cabbage indicator and reference set.
7. Add 5 drops of 0.1 M NaOH to each test tube and record the pH.
8. Add 5 more drops of 0.1 M NaOH to each test tube and record the pH.
9. Determine the change in pH for each solution.
10. Identify which solutions are buffered.

Reference Colors for Red Cabbage Indicator

pH	Color	pH	Color
1		8	
2		9	
3		10	
4		11	
5		12	
6		13	
pH 7			

Part A: Measuring pH

Substance	Color of Indicator	pH	pH Using Meter	Acidic or Basic	
Vinegar				Acidic	Basic
Ammonia				Acidic	Basic
Lemon juice				Acidic	Basic
Soda				Acidic	Basic
Laundry detergent				Acidic	Basic
Liquid hand soap				Acidic	Basic
Antacid				Acidic	Basic
Aspirin				Acidic	Basic

Part B: Buffers

Adding acid to solutions:

Solution	Initial pH	pH after Five Drops HCl	pH after Additional Five Drops HCl	pH Change	Buffer	
H_2O					Yes	No
0.1 NaCl					Yes	No
High buffer pH					Yes	No
Low buffer pH					Yes	No

Adding base to solutions:

Solution	Initial pH	pH after Five Drops NaOH	pH after Additional Five Drops NaOH	pH Change	Buffer	
H_2O					Yes	No
0.1 NaCl					Yes	No
High buffer pH					Yes	No
Low buffer pH					Yes	No

Questions and Problems

1. A solution has a $[OH^-] = 4.0 \times 10^{-5}$. What is the concentration of the hydronium ion, $[H_3O^+]$ and the pH of the solution? (show your work).

2. The following shows the dissociation of HCl in water:

$$HCl_{(aq)} + H_2O_{(l)} \xrightarrow{\text{Yields}} H_3O^+_{(aq)} + Cl^-_{(aq)}$$

   a. You mix up a sample that has 0.0084 moles of HCl in enough water to make a 1500 mL solution. Calculate the molarity of this solution. (show your work).

   b. What is the concentration of the hydronium ion $[H_3O^+]$?

   c. What is the pH of this solution? (show your work)

   d. Is this an acidic or basic solution?

3. In the buffered solutions portion of the lab, which solutions showed the greatest change in pH? Explain.

4. Which solutions showed the least change in pH? Explain.

5. Normal pH for the human body is in a narrow range, 7.35 to 7.45. A patient with acidotic blood, pH 7.3, may be treated using an alkali such as sodium carbonate. Why would this treatment raise the pH of the blood?

LAB 14

Organic Compounds: Alkanes

Organic compounds are typically differentiated from inorganic compounds by their (1) physical and chemical properties, (2) solubility, and (3) combustibility (flammability). We will briefly examine each of these differences in more detail.

1. Physical and chemical properties: Organic compounds are composed chiefly of carbon and hydrogen, and sometimes oxygen, sulfur, nitrogen, or one or more of the halogens (F, Cl, Br, or I). Covalent bonds in organic compounds versus ionic bonds in inorganic compounds account for most of the differences in physical and chemical properties, such as melting points and boiling points; that is, organic compounds tend to have much lower melting and boiling points than inorganic compounds.
2. Solubility: Ionic inorganic compounds are typically soluble in water, a polar solvent, and insoluble in nonpolar organic solvents, whereas nonpolar organic compounds are insoluble in water but are soluble in nonpolar organic solvents, leading to the general rule that should be remembered: "Like dissolves like."
3. Combustibility (flammability): Inorganic compounds generally do not react with oxygen to undergo the reaction known as combustion, while organic compounds do, forming carbon dioxide, water, and releasing heat (thermal) energy. This exothermic reaction occurs when gasoline burns with oxygen in the engine of a car or when methane, also known as natural gas, burns in a heater or stove. The balanced equations for the combustion of methane can be written as follows:

$$CH_4\ (g) + 2O_2\ (g) \rightarrow CO_2\ (g) + 2H_2O\ (g) + \text{heat}$$

Hydrocarbons, as their name implies, are molecules composed solely of hydrogen and carbon atoms. Alkanes and their cyclic forms, known as **cycloalkanes**, represent *saturated* hydrocarbons since their carbon atoms are connected by only single bonds and their molecules contain the maximum number of hydrogens allowable with carbon's ability to form four total bonds. An **expanded structural formula** shows all of the atoms and the bonds connected to each atom:

Ethane

cyclopropane

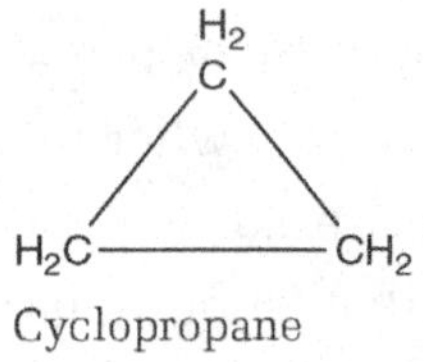

Cyclopropane
Source: Kendall Hunting Publishing

While a **condensed structural formula** shows the carbon atoms each grouped with the attached number of hydrogen atoms:

CH_3–CH_3 ethane

A cycloalkane can also be written as what is known as a **skeletal formula**:

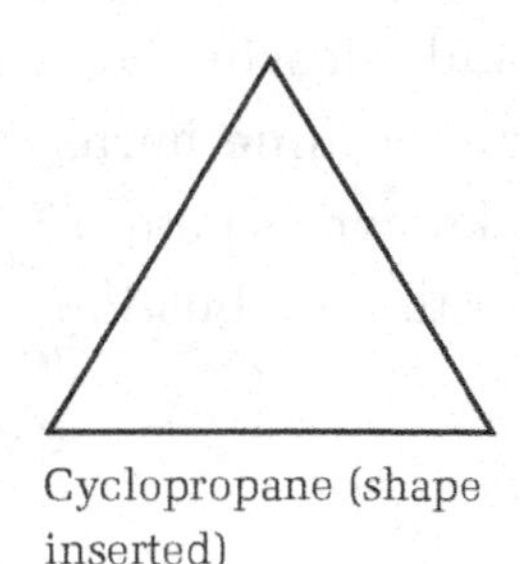
Cyclopropane (shape inserted)

Isomers represent two or more different structural formulas for the same molecular formula. One isomer cannot be converted to the other without breaking and forming new bonds. Isomers also have different physical and chemical properties. This phenomenon of isomerism is one of the reasons for the vast number of organic compounds.

Haloalkanes consist of one or more halogen atoms, such as chlorine (Cl) or bromine (Br), that replace one or more hydrogen atoms of an alkane or cycloalkane.

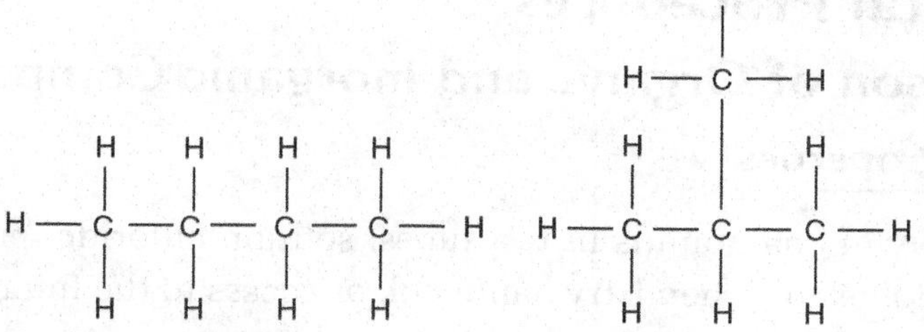

butane (C_4H_{10}) 2-methylpropane (C_4H_{10})

Isomers of C_4H_{10}

Source: Kendall Hunting Publishing

Expanded Structural Formula	Condensed Structural Formula	Name
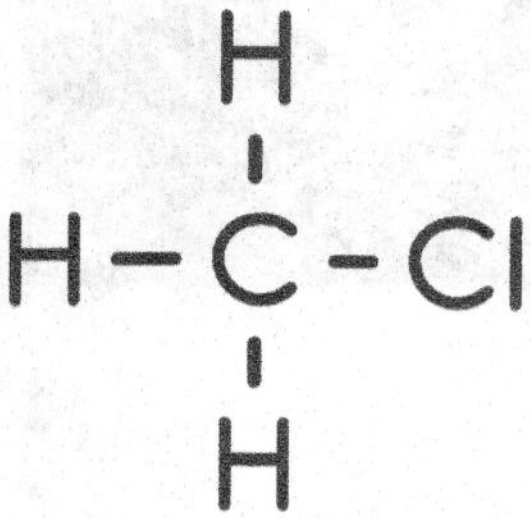	CH_3—Cl	Chloromethane (Methyl chloride)
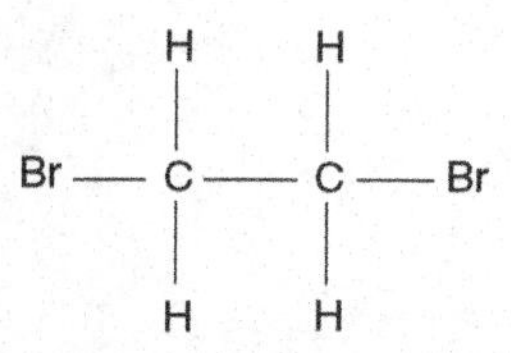	Br–CH_2–CH_2–Br	1,2-Dibromoethane

Source: Kendall Hunting Publishing

Experimental Procedures

A. Comparison of Organic and Inorganic Compounds

A.1. Physical Properties

Materials: Display of compounds in test tubes: sodium chloride, potassium iodide, toluene, and cyclohexane; chemistry handbook or access to the Internet.

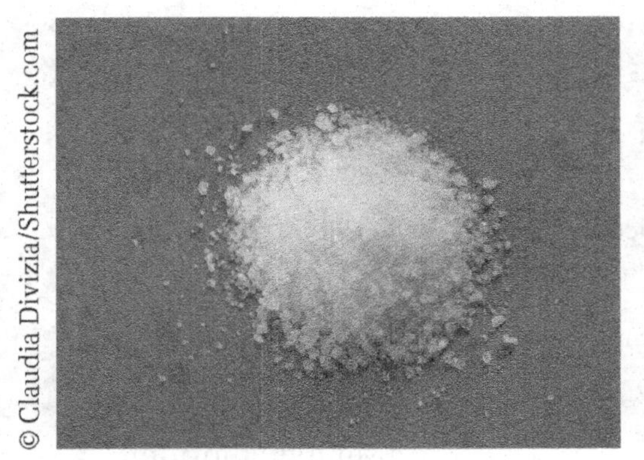

Sodium Chloride

Potassium Iodide

Toluene and Cyclohexane

1. Observe the samples in the test tubes: sodium chloride, potassium iodide, toluene, and cyclohexane. Record the molecular formula for each, using a reference source if necessary.
2. Record the physical state (solid, liquid, or gas).
3. Look up the melting point of each compound using a chemistry handbook or the Internet and record.
4. Identify the types of bonds in each as either ionic or covalent.
5. Identify each as either an organic or inorganic compound.

A.2. Solubility

Materials: Display of samples containing NaCl added to cyclohexane and water, and toluene added to cyclohexane and water.

1. Observe the samples to which NaCl is added to the nonpolar solvent cyclohexane and record whether NaCl is soluble (S) or insoluble (I) in cyclohexane.
2. Observe the sample in which NaCl is added to the polar solvent water and record whether NaCl is soluble (S) or insoluble (I) in water.
3. Identify NaCl as organic or inorganic.
4. Observe the sample in which toluene is added to the nonpolar solvent cyclohexane and record whether toluene is soluble (S) or insoluble (I) in cyclohexane.
5. Observe the sample in which toluene is added to the polar solvent water and record whether toluene is soluble (S) or insoluble (I) in water,
6. Identify toluene as organic or inorganic.

A.3. Combustion

Materials: Two evaporating dishes, spatulas, wood splints, NaCl(s), and cyclohexane(l) (demonstration).

1. Observe the instructor ignite a wooden splint and hold the flame to a sample of NaCl. Record whether NaCl undergoes combustion and identify NaCl as an organic or inorganic compound.
2. Observe the instructor ignite a wooden splint and hold the flame to a sample of cyclohexane. Record whether cyclohexane undergoes combustion and identify cyclohexane as an organic or inorganic compound.

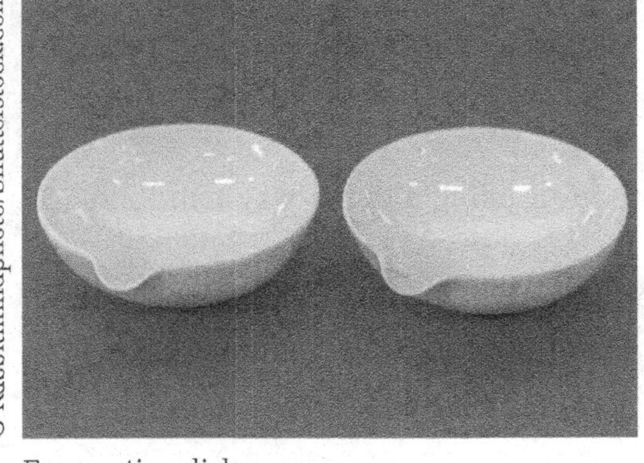

Evaporating dishes

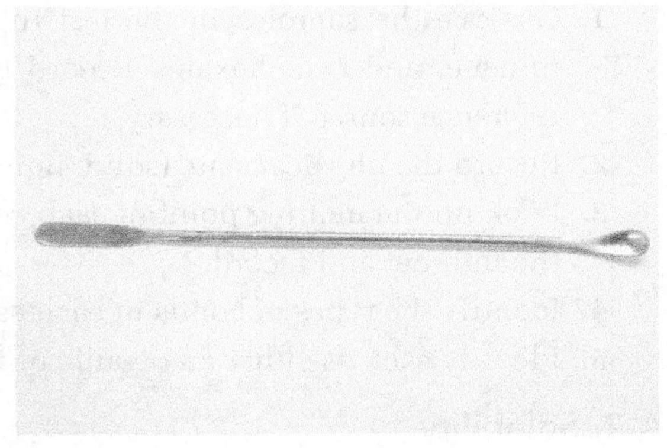

Spatula

Wooden splint

B. Alkanes

B.1. Structures and Names of Alkanes

Materials: Display of models of methane, ethane, and propane.

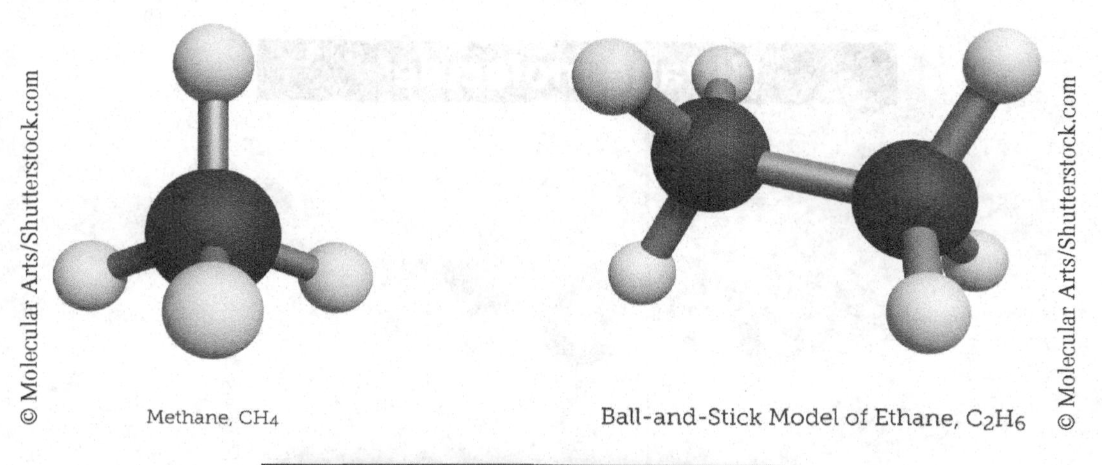

Methane, CH_4

Ball-and-Stick Model of Ethane, C_2H_6

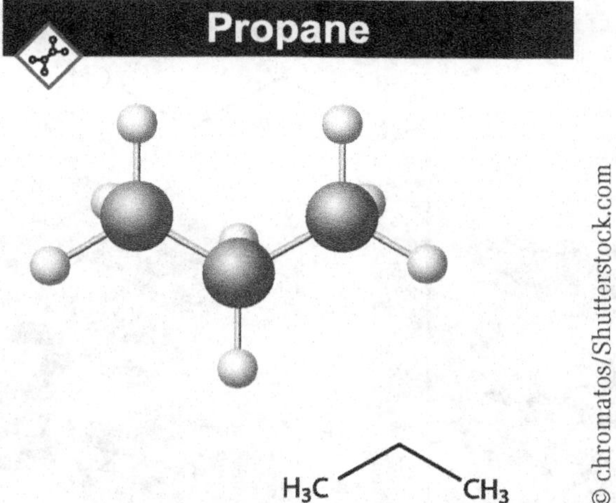

1. Observe the model of methane and draw its expanded and condensed structural formulas.
2. Observe the model of ethane and draw its expanded and condensed structural formulas.
3. Observe the model of propane and draw its expanded and condensed structural formulas.

B.2. Isomers of Alkanes

Materials: Display of models of butane, 2-methylpropane, 2,2-dimethylpropane; chemistry handbook or access to the Internet.

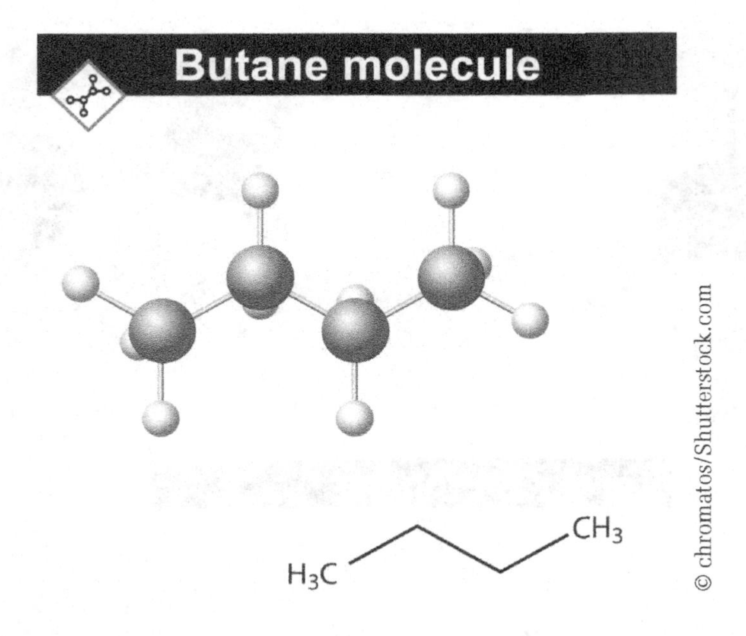

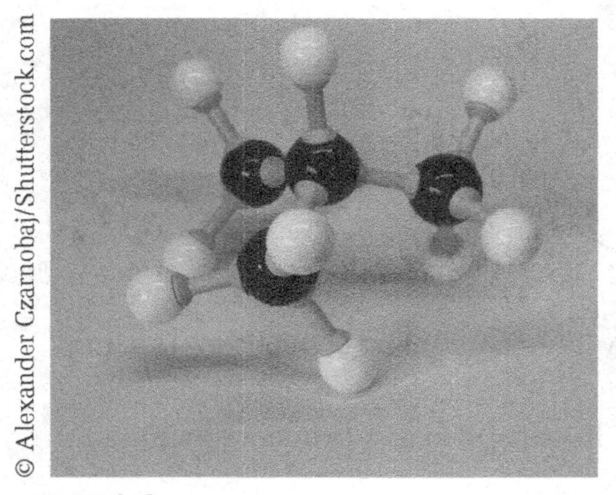

2-Methylpropane

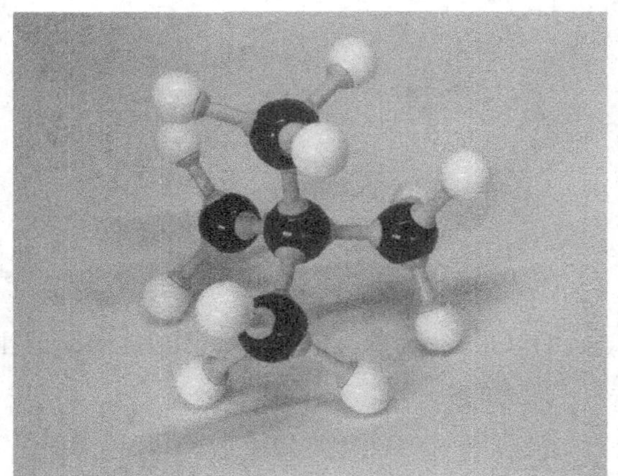

2,2-Dimethylpropane

1. Observe the model of butane and draw its expanded and condensed structural formulas.
2. Observe the model of 2-methylpropane, an isomer of butane, and draw its expanded and condensed structural formulas.
3. Using a chemistry handbook or the Internet, obtain the molar mass, boiling point, and density of each isomer.
4. Draw the condensed structural formulas for the three isomers of C_5H_{12}: pentane, 2-methylbutane, and 2,2-dimethylpropane.
5. Using a chemistry handbook or the Internet, obtain the molar mass, boiling point, and density of each isomer.

B.3. Cycloalkanes

Materials: Display of models of cyclopentane and cyclohexane.

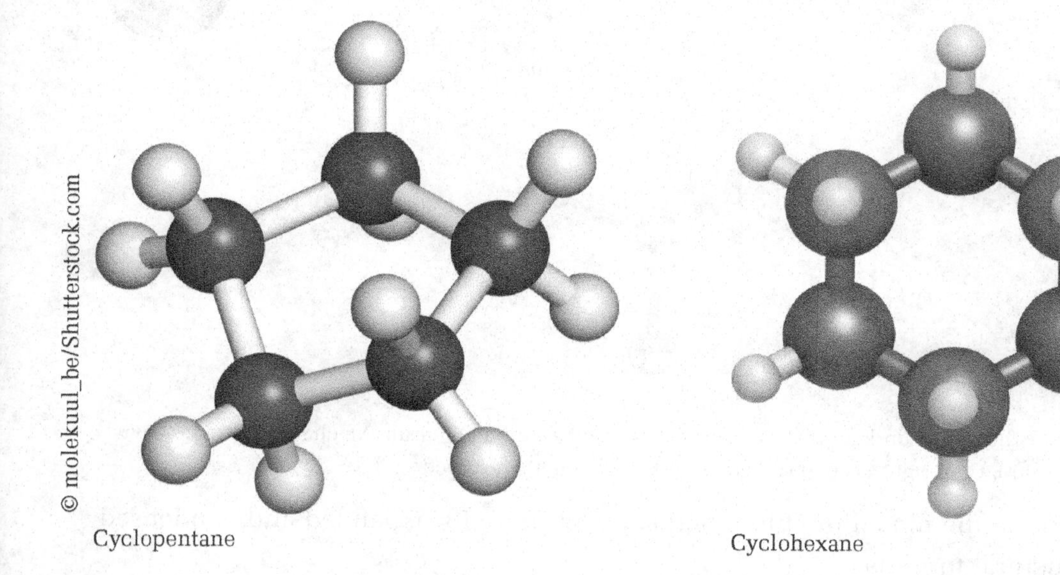

Cyclopentane

Cyclohexane

1. Observe the model of cyclopentane and draw its condensed structural and skeletal formulas.
2. Observe the model of cyclohexane and draw its condensed structural and skeletal formulas.

B.4. Haloalkanes

Materials: Display of models of chloromethane: 1,2-dibromoethane and 2-iodopropane.

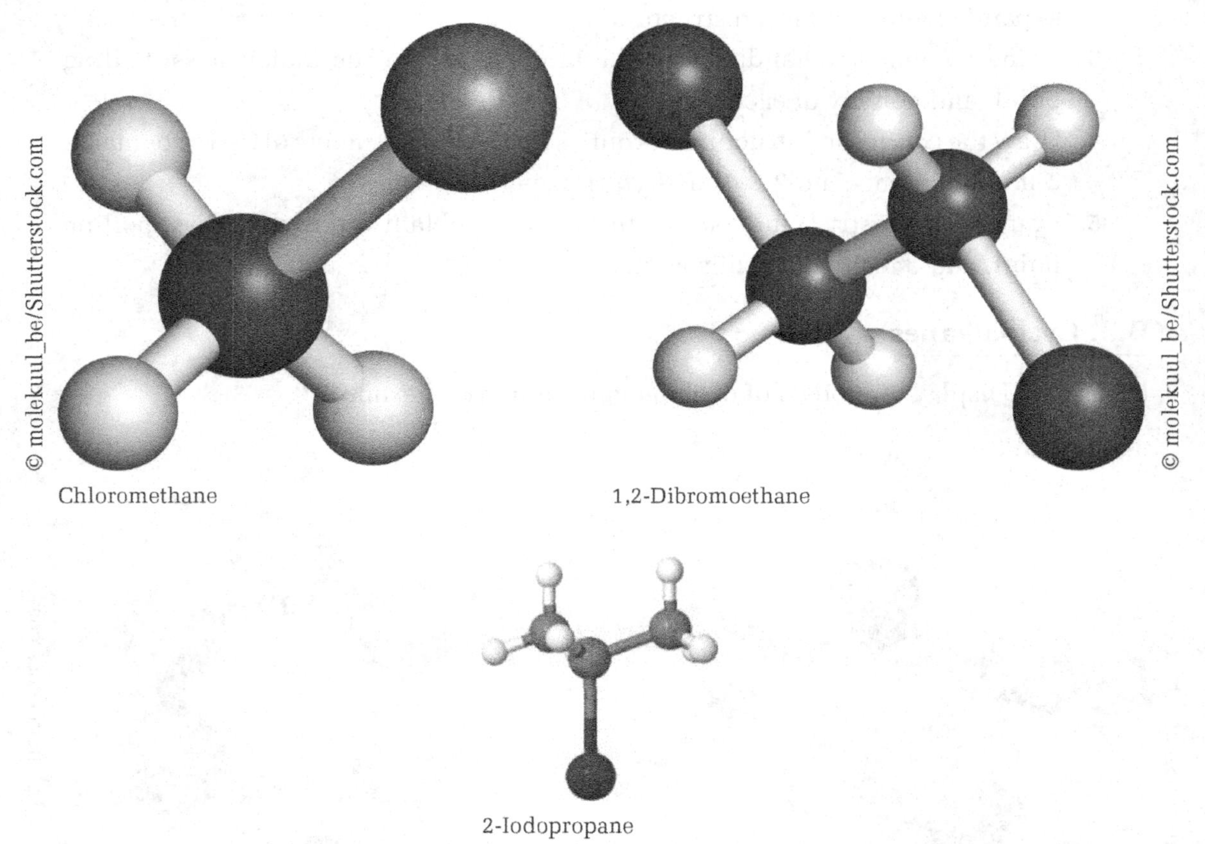

Chloromethane

1,2-Dibromoethane

2-Iodopropane

1. Observe the model of chloromethane and draw its expanded and condensed structural formulas.
2. Observe the model of 1,2-dibromoethane and draw its expanded and condensed structural formulas.
3. Observe the model of 2-iodopropane and draw its expanded and condensed structural formulas.
4. Draw the condensed structural formulas of the four isomers of $C_3H_6Cl_2$ and name each isomer.

Date ________________ Name ________________________________

Section ________________ Team ________________________________

Instructor ________________ ________________________________

Prelab Study Questions

1. What elements are present in alkane?

2. Would you expect hexane to be soluble in water? Why or why not?

3. Which is more flammable, hexane or potassium sulfate?

4. Draw the condensed structural formula of hexane.

5. Why are 1-chlorobutane and 2-chlorobutane structural isomer?

Date ________________ Name ________________________________

Section ________________ Team ________________________________

Instructor ________________ ________________________________

Report Sheet for Lab 14: Organic Compounds: Alkanes

A. Comparison of Organic and Inorganic Compounds

A.1. Physical Properties

Name	Formula	Physical State	Melting Point	Type of Bonds	Organic or Inorganic
Sodium chloride					
Potassium iodide					
Toluene					
Cyclohexane					

A.2. Solubility

Solute	Soluble in Cyclohexane?	Soluble in Water?	Organic or Inorganic?
NaCl			

Solute	Soluble in Cyclohexane?	Soluble in Water?	Organic or Inorganic?
Toluene			

A.3. Combustion

Compound	Undergoes Combustion? (Yes/No)	Organic or Inorganic?
NaCl		
Toluene		

Questions and Problems

1. From your observations of the chemical and physical properties of NaCl and Toluene as either organic or inorganic compounds, complete the following table:

Property	Organic Compounds	Inorganic Compounds
Elements		
Bonding (covalent or ionic)		
Melting point (low or high)		
Solubility in water (I or S)		
Flammability (yes or no)		

2. A white solid is solid in water and is not flammable. Would you expect it to be organic or inorganic and explain your reason?

3. A clear liquid forms a layer when added to water. Would you expect it to be organic or inorganic and explain your reason?

B. Alkanes

B.1. Structures and Names of Alkanes

Compound	Expanded Structural Formula	Condensed Structural Formula
Methane		
Ethane		
Propane		

B.2. Isomers of Alkanes

Isomers of C_4H_{10}

Compound	Butane	2-Methylpropane
Expanded structural formula		
Condensed structural formula		
Molar mass		
Boiling point		
Density		

Isomers of C_5H_{12}

Compound	Pentane	2-Methylbutane	2,2-Dimethylpropane
Condensed structural formula			
Molar mass			
Boiling point			
Density			

Questions and Problems

4. What physical property is identical for the isomers of C_5H_{12}?

B.3. Cycloalkanes

Compound	Condensed Structural Formula	Skeletal Formula
Cyclopentane		
Cyclohexane		

B.4. Haloalkanes

Compound	Expanded Structural Formula	Condensed Structural Formula
Chloromethane		
1,2-Dibromoethane		
2-Iodopropane		

Isomers of $C_3H_6Cl_2$

Condensed Structural Formula	Condensed Structural Formula
Name:	Name:
Name:	Name:

LAB 15

Reactions of Unsaturated Hydrocarbons

While alkanes are *saturated* hydrocarbons containing single bonds between carbon atoms, alkenes and alkynes are *unsaturated* hydrocarbons containing double or triple bonds between carbon atoms, respectively. These multiple unsaturated bonds are very reactive sites. Aromatic compounds are hydrocarbons with a benzene ring. International Union of Pure and Applied Chemistry (IUPAC) names for alkenes and alkynes relate to the names of alkanes with the same number of carbon atoms, but ending in *ene* or *yne*, respectively (see Table 15.1). Skeletal formulas are written for unsaturated hydrocarbons to show the position of the double or triple bond.

Table 15.1 Comparison of Names for Alkanes, Alkenes, and Alkynes

Alkane	Alkene	Alkyne
CH_3–CH_3 Ethane	H_2C=CH_3 Ethene (ethylene)	HC≡CH Ethyne (acetylene)
CH_3–CH_2–CH_3 Propane	Propene	CH_3–C≡CH Propyne

Source: Bruce Kowiatek

When bromine (Br_2) reacts with an alkene, the dark red color of the Br_2 disappears quickly due to the bromine atoms forming bonds with the carbon atoms of the double bond. Rapid disappearance of the red color, therefore, indicates that the compound contains an unsaturated site.

$$H_2C{=}CH_2 + Br_2 \longrightarrow H{-}CHBr{-}CHBr{-}H$$

Ethylene | Bromine (brownish-red) | 1, 2-Dibromoethane (colorless)

Source: Bruce Kowiatek

The addition reaction of bromine with alkanes is much slower and requires light. In this reaction, one H atom in the alkane is replaced with a Br atom. When bromine is added to an alkane, the red color persists for several minutes before it fades. Aromatic compounds such as benzene rings are unreactive with bromine.

$$CH_3\text{–}CH_3 + Br_2 \xrightarrow{\text{Light}} CH_3\text{–}CH_2\text{–}Br + HBr_{(g)}$$

Colorless *Red* (slowly) *Colorless* *Pungent odor*

benzene + $Br_2 \rightarrow$ No Reaction

Colorless *Red*

In the potassium permanganate ($KMnO_4$) oxidation test, alkenes, but not alkanes or aromatic compounds, are oxidized. The purple color of $KMnO_4$ changes to the muddy brown of manganese dioxide (MnO_2) in this reaction, the product being a diol:

$$H_2C{=}CH_2 + KMnO_4 \rightarrow HO\text{–}CH_2\text{–}CH_2\text{–}OH + MnO_2$$

Colorless *Purple* *Colorless* *Brown*

Your test results will be used to identify your unknown as one of the experimental compounds.

Experimental Procedures

A. Types of Unsaturated Hydrocarbons

Materials: Display of models of the unsaturated hydrocarbons ethene (ethylene), propene, cyclobutene, *cis*-2-butene, and ethyne.

Observe the models and draw each of their condensed structures.

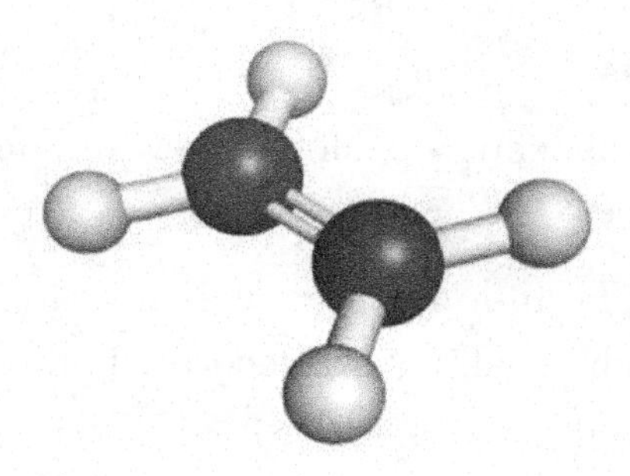

Ethene
Source: Bruce Kowiatek

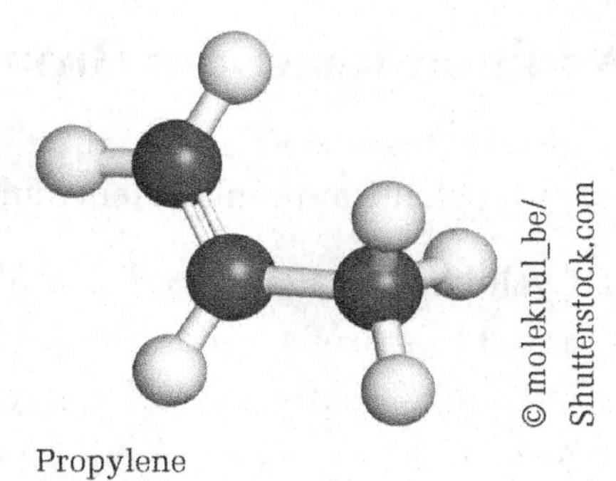

Propylene

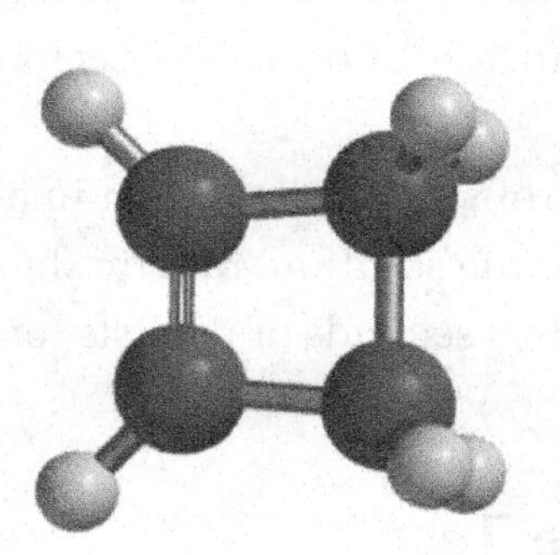

Cyclobutene
Source: Bruce Kowiatek

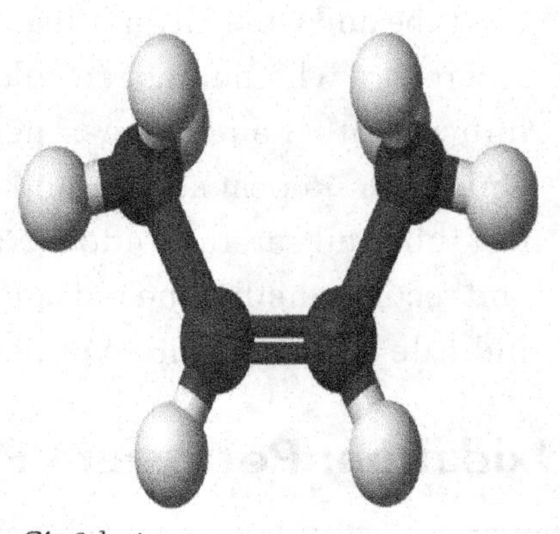

Cis-2-butene
Source: Bruce Kowiatek

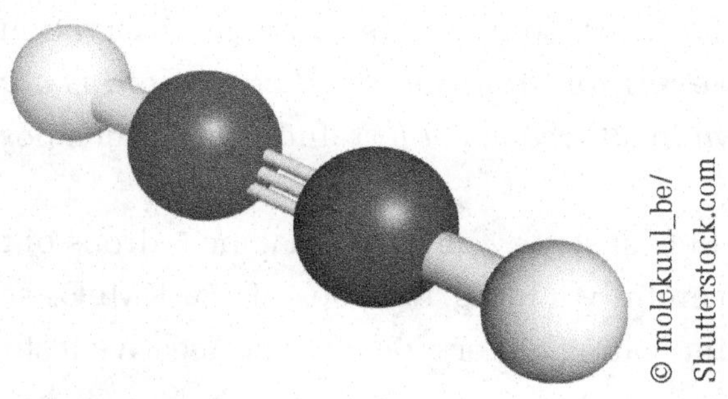

Ethyne

B. Addition Reaction: Bromine Test

Materials: 4 test tubes, test tube rack, cyclohexane, cyclohexene, unknown, 1% bromine solution in methylene chloride.

1. **Cyclohexene.** Place 15 drops of cyclohexene in a dry test tube and carefully add 3 to 4 drops of bromine solution. Mix by shaking and record whether the red color disappears in 2 to 3 seconds or persists longer.
2. *Note: a positive test for an unsaturated compound is a fast disappearance of the red bromine color.* Indicate either a positive or negative test result.

   Cyclohexane. Repeat steps 1 and 2 by placing 15 drops of cyclohexane in a dry test tube and carefully adding 3 to 4 drops of bromine solution. Mix by shaking and record whether the red color disappears in 2 to 3 seconds or persists longer: indicate either a positive or negative test result.

   Unknown. Repeat steps 1 and 2 by placing 15 drops of the unknown in a dry test tube and carefully adding 3 to 4 drops of bromine solution. Mix by shaking and record whether the red color disappears in 2 to 3 seconds or persists longer; indicate either a positive or negative test result.

C. Oxidation: Potassium Permanganate Test

Materials: 4 test tubes, test tube rack, 1% $KMnO_4$, cyclohexane, cyclohexene, unknown.

1. **Cyclohexene.** Place 5 drops of cyclohexene in a dry test tube and carefully add 15 drops of 1% $KMnO_4$ solution. Record your observations.
2. *Note: a positive test for an unsaturated compound is a change in color from purple to brown in 60 seconds or less.* Indicate either a positive or negative test result.

   Cyclohexane. Repeat steps 1 and 2 by placing 5 drops of cyclohexane in a dry test tube and carefully adding 15 drops of 1% $KMnO_4$ solution. Record your observations, indicating either a positive or negative test result.

Unknown. Repeat steps 1 and 2 by placing 5 drops of the unknown in a dry test tube and carefully adding 15 drops of 1% $KMnO_4$ solution. Record your observations, indicating either a positive or negative test result.

D. Identification of an Unknown

From your test results, indicate whether the bromine test and $KMnO_4$ test were positive or negative; identify the unknown as a saturated or unsaturated hydrocarbon; and state your reasoning.

Date ________________ Name ________________________________

Section ________________ Team ________________________________

Instructor ________________ ________________________________

Prelab Study Questions

1. **A.** What color change(s) occur when bromine reacts with an alkene?

   B. What color change(s) take place when $KMnO_4$ reacts with an alkene?

2. Why is the reaction of ethene with bromine called an addition reaction?

3. If a bromine test were used to distinguish between hexane and 2-hexene, in which would the red color of bromine persist?

4. If a permanganate test were used to distinguish between hexane and 2-hexene, in which would the purple color of $KMnO_4$ persist?

5. Draw the condensed structural formula for each of the following:

   a. 2-Methyl-1-pentene

   b. 3-Methylcyclopentene

Date ________________ Name ________________________________

Section ________________ Team ________________________________

Instructor ________________ ________________________________

Report Sheet for Lab 15: Reactions of Unsaturated Hydrocarbons

A. Types of Unsaturated Hydrocarbons

Models of Unsaturated Hydrocarbons

Compound	Condensed Structural Formula	Compound	Condensed Structural Formula
Ethene		*Cis*-2-butene	
Propene		Ethyne (acetylene)	
Cyclobutene			

B. Addition Reactions: Bromine Test

Compound	Disappearance of Bromine Color (Fast/Slow)	Test Results (Positive/Negative)
Cyclohexene		
Cyclohexane		
Unknown		

C. Oxidation: Potassium Permanganate Test

Compound	$KMnO_4$ Observations	Test Results (Positive/Negative)_
Cyclohexene		
Cyclohexane		
Unknown		

D. Identification of an Unknown

Unknown Number	Bromine Test	$KMnO_4$ Test	Saturated or Unsaturated

LAB 16

Alcohols and Phenols

Organic compounds containing a hydroxyl group (–OH) are known as alcohols. Methanol is the simplest alcohol. Ethanol, found in alcoholic beverages and preservatives, is used as a solvent, and 2-propanol, also known as rubbing or isopropyl alcohol, can be found in astringents and perfumes.

A. Structures of Alcohols and Phenol

CH_3–OH	CH_3–CH_2–OH	CH_3–CHOH–CH_3
Methanol	Ethanol	2-Propanol
(Methyl alcohol)	(Ethyl alcohol)	(Isopropyl alcohol)

Phenol is a benzene ring with a hydroxyl group. Concentrated solutions of phenol are caustic, causing burns; however, derivatives of phenol like thymol are used as antiseptics and can sometimes be found in cough drops and mouthwashes.

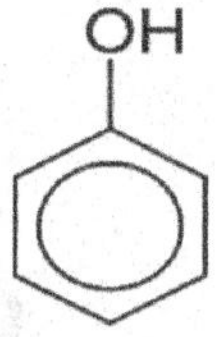

OH

Phenol

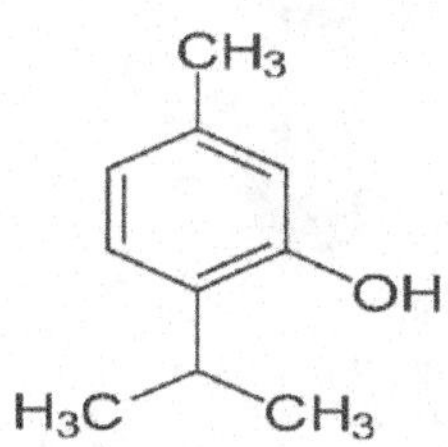

CH_3 OH H_3C CH_3

Thymol
(2-Isopropyl-5-
methylphenol)
Source: Bruce
Kowiatek

Classification of Alcohols

In primary (1°) alcohols, the carbon atom attached to the –OH group is bonded to one other carbon atom. In secondary (2°) alcohols, the carbon with the –OH is attached to two carbon atoms, and in tertiary (3°) alcohols, it is attached to three carbon atoms.

```
    H   H
    |   |
H — C — C — OH
    |   |
    H   H
```

Ethanol
Primary (1°) alcohol
Source: Bruce Kowiatek

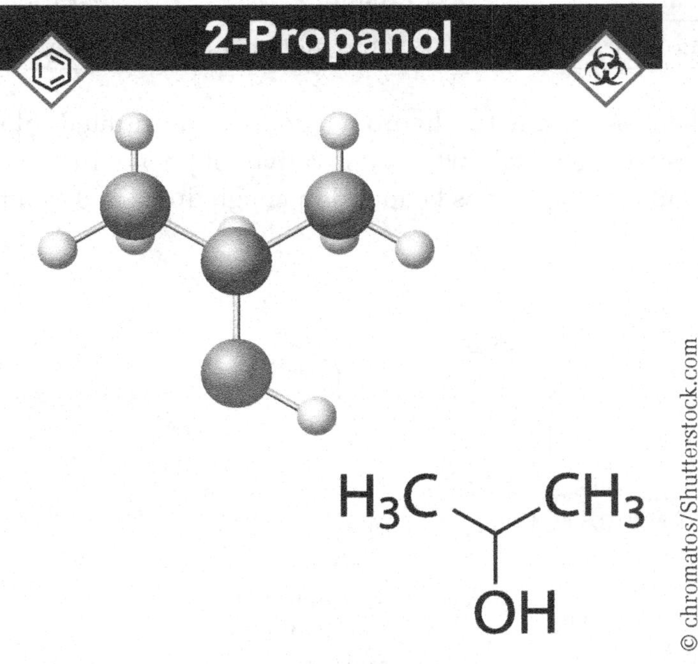

2-Propanol
Secondary (2°) alcohol

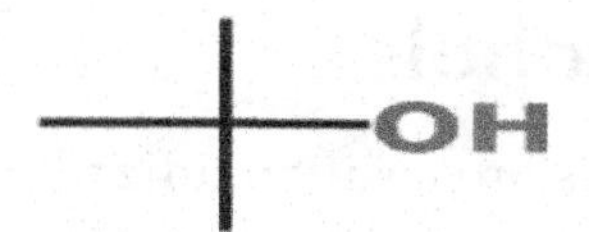

2-Methyl-2-Propanol
Tertiary (3°) alcohol
Source: Bruce Kowiatek

B. Properties of Alcohols and Phenol

The polarity of the hydroxyl group (–OH) makes alcohols with one to three carbons completely soluble in water since they can form hydrogen bonds. Alcohols with four carbons are somewhat soluble, while the large hydrocarbon portion in longer chain alcohols makes them insoluble in water.

Hydrogenbond

109°

105°

Ethyl alcohol in water
(hydrogen bonds form
between the hydroxyl groups
and H and O atoms in water.
Source: Bruce Kowiatek

Acidity of Phenol

Phenol acts as a weak acid in water because the hydroxyl group ionizes slightly; although phenol has six carbon atoms, the polarity of the hydroxyl group makes it soluble in water.

$$C_6H_5OH + H_2O \rightleftharpoons C_6H_5O^- + H_3O^+$$

a phenoxide ion

Source: Bruce Kowiatek

C. Oxidation of Alcohols

Primary and secondary alcohols are easily oxidized; an oxidation consists of removing an H atom from the –OH group and another H atom from the C atom attached to the –OH group. Tertiary alcohols do not oxidize since there are no H atoms on that C atom attached to the –OH. Primary and secondary alcohols may be distinguished from tertiary ones using a chromate (CrO_4^{2-}) solution; the orange color of the chromate solution turns green when an oxidation has occurred.

$$CH_3\text{–}CH_2\text{–}OH + CrO_4^{2-} \xrightarrow{H^+} CH_3\text{–}CHO + H_2O + Cr^{3+}$$

1° alcohol Orange Aldehyde Water Green

$$CH_3\text{–}CH(CH_3)\text{–}OH + CrO_4^{2-} \xrightarrow{H^+} CH_3\text{–}CO\text{–}CH_3 + H_2O + Cr^{3+}$$

2° alcohol Orange Ketone Water Green

$$CH_3\text{–}C(CH_3)_2\text{–}OH + CrO_4^{2-} \xrightarrow{H^+} \text{No Reaction (remains orange)}$$

3° alcohol Orange

D. Iron(III) Chloride Test

Phenols react with the Fe^{3+} ion in an iron(III) chloride ($FeCl_3$) solution to give complex ions with strong colors from red to purple.

Phenol + $Fe^{3+} \rightarrow Fe^{3+}$ (phenol complex)

Colorless Yellow Purple

E. Identification of an Unknown

The group of tests for alcohols and phenols described above may be used to identify the functional group and family of an unknown.

Experimental Procedures

A. Structures of Alcohols and Phenols

Materials: Display of models: ethanol, 2-propanol, and 2-methyl-2-propanol

Observe the models and draw the condensed structural formula of each model in the display as well as for phenol; classify each as a primary, secondary, or tertiary alcohol.

B. Properties of Alcohols and Phenol

Materials: 6 test tubes, 6 rubber stoppers for the test tubes, pH paper, stirring rod, ethanol, 2-propanol, 2-methyl-2-propanol (*t*-butyl alcohol), cyclohexanol, 20% phenol, and an unknown. ***Caution: Avoid contact with phenol.***

pH

1. Place 10 drops of ethanol, 2-propanol, *t*-butyl alcohol (2-methyl-2-propanol), cyclohexanol, 20% phenol, and the unknown into 6 separate test tubes. Obtain pH paper and use a stirring rod to place a drop of each on the pH paper, cleaning the stirring rod between applications, and comparing the pH paper color with the color chart of the container to determine the pH. Record your observations and save the test tubes and alcohols for step 2.

   Solubility in water

2. Add 2 mL of water (40 drops) to each test tube, place rubber stoppers over each, shake and determine the solubility of each alcohol in water. If the compound is water soluble, a clear solution with no separate layers will result; if insoluble, either a cloudy mixture or separate layers will form. Record your observations.

   DISPOSE OF ORGANIC COMPOUNDS IN DESIGNATED WASTE CONTAINERS!

C. Oxidation of Alcohols

Materials: 6 test tubes, stirring rod, ethanol, 2-methyl-2-propanol (*t*-butyl alcohol), cyclohexanol, 20% phenol, unknown, 2% chromate solution, beaker of ice-cold water.

1. Place 8 drops of ethanol, 2-propanol, 2-methyl-2-propanol (*t*-butyl alcohol), cyclohexanol, 20% phenol and the unknown into six separate test tubes. Carefully add 2 drops of chromate solution to each and stir carefully with a stirring rod to allow the alcohol to react. ***Caution: Chromate solution contains concentrated H_2SO_4, which is corrosive.*** Look for any color change in the chromate solution. If the test tube becomes hot, place in a beaker of ice-cold water. Record your observations of the color after 2 minutes.
2. Draw the condensed structural formula of each alcohol.
3. Classify each alcohol as primary (1°), secondary (2°), or tertiary (3°).
4. If reaction have occurred, draw the condensed structural formulas of the products; if no change of color was observed, then no oxidation took place and, in this case, write NR for "no reaction."

DISPOSE OF ORGANIC COMPOUNDS IN DESIGNATED WASTE CONTAINERS!

D. Iron(III) Chloride Test

Materials: 6 test tubes, stirring rod, ethanol, 2-methyl-2-propanol (*t*-butyl alcohol), cyclohexanol, 20% phenol, unknown, 1% $FeCl_3$ solution.
Place 5 drops of the alcohols and unknown in separate test tubes and add 5 drops of 1% $FeCl_3$ solution to each; stir and record observations.

DISPOSE OF ORGANIC COMPOUNDS IN DESIGNATED WASTE CONTAINERS!

E. Identification of an Unknown

Use the test results to identify your unknown as one of the five compounds used in this experiment; draw the condensed structural formula of the unknown.

Date ________________ Name ________________________________

Section ________________ Team ________________________________

Instructor ________________ ________________________________

Prelab Study Questions

1. What is the functional group of an alcohol and a phenol?

2. Why are some alcohols soluble in water?

3. Classify each of the alcohols as primary, secondary, or tertiary.
   a. 3-Pentanol ____________
   b. 2-Methyl-2-butanol ____________
   c. 1-Propanol ____________

4. If you add the oxidizing agent chromate to each of the following, would a green Cr^{3+} solution result?
   a. 3-Pentanol ____________
   b. 2-Methyl-2-butanol ____________
   c. 1-Propanol ____________

5. If an alcohol has a pH of 5, would it be a primary, secondary, or tertiary alcohol, or would it be a phenol?

Date ________________ Name ______________________________

Section ________________ Team ______________________________

Instructor ________________ ______________________________

Report Sheet for Lab 16: Alcohols and Phenols

A. Structures of Alcohols and Phenols

Ethanol	2-Propanol
Classification:	
2-Methyl-2-propanol (*t*-butyl alcohol)	Phenol
Classification:	/// //////////////////////

Questions and Problems

1. Draw the condensed structural formula and give the classification for each of the following alcohols:

1-Pentanol	3-Pentanol
Classification:	
Cyclopentanol	1-Methylcyclopentanol
Classification:	

B. Properties of Alcohols and Phenols

Unknown number ________

Alcohol	1. pH	2. Soluble in Water?
Ethanol		
2-Propanol		
2-Methyl-2-propanol		
Cyclohexanol		
Phenol		
Unknown		

C. Oxidation of Alcohols

Alcohol	1. Color after 2 Minutes	2. Condensed Structural Formula	3. Classification	4. Condensed Structural Formula of Oxidation Product
Ethanol				
2-Propanol				
2-Methyl-2-propanol				
Cyclohexanol				
Phenol				
Unknown				

2. Draw the condensed structural formula of the product of the following reactions (if no reaction, write NR):

a. $CH_3–CH_2–CH_2–OH \xrightarrow{[O]}$

b. $CH_3–CHOH–CH_2–CH_3 \xrightarrow{[O]}$

[O]

c.

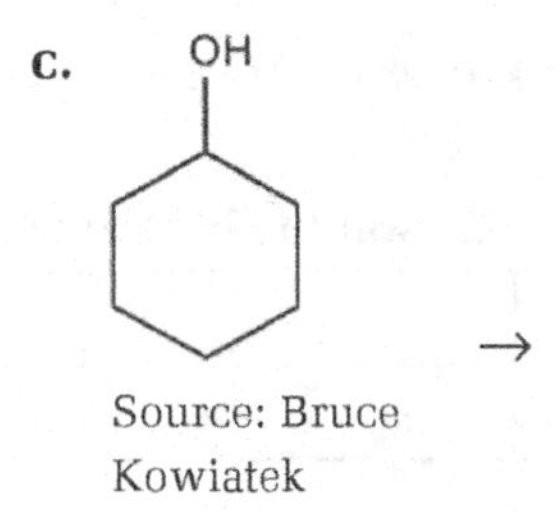

Source: Bruce Kowiatek

D. Iron(III) Chloride Test

Alcohol	$FeCl_3$ Test (Color)
Ethanol	
2-Propanol	
t-Butyl alcohol	
Cyclohexanol	
Phenol	
Unknown	

E. Identification of an Unknown

Unknown Number ____________		
Summary of Testing	Test Result	Conclusion
B.1 pH		
B.2 Soluble in water?		
C.1 CrO_4^{2-}		
D $FeCl_3$		
Name of Unknown	**Condensed Structural Formula**	

LAB 17

Aldehydes and Ketones

Aldehydes and ketones both contain what is known as a carbonyl group. In aldehydes, the carbonyl group has a hydrogen atom attached; this aldehyde functional group occurs at the end of carbon chains, while the ketone functional group has the carbonyl located between two of the carbon atoms within the chain.

A. Structures of Some Aldehydes and Ketones

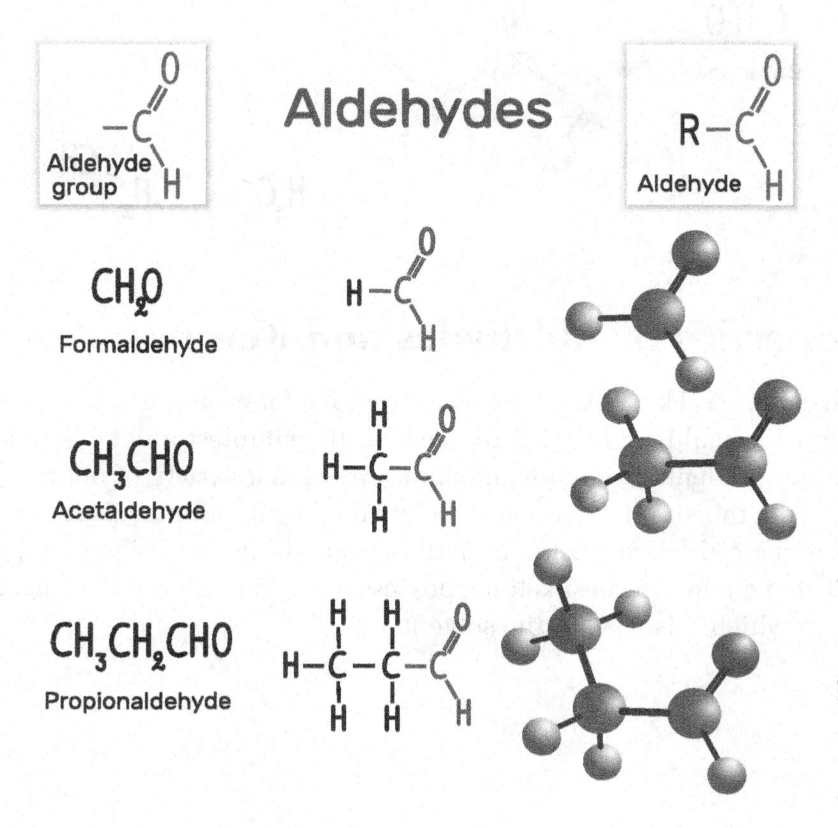

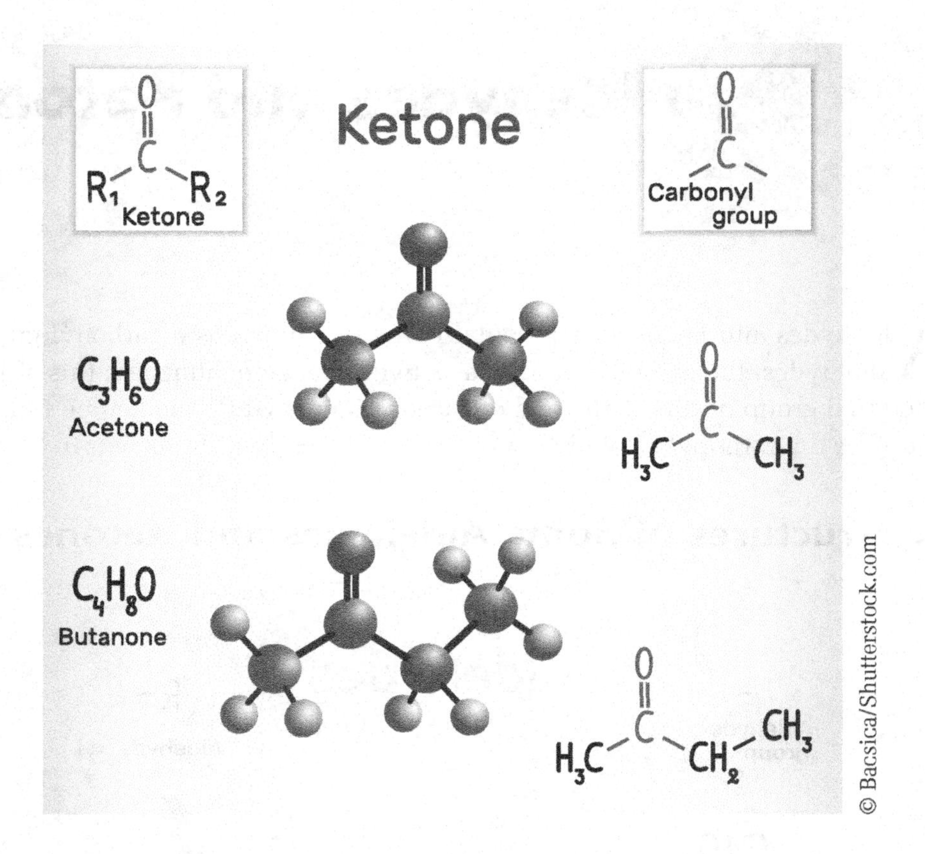

B. Properties of Aldehydes and Ketones

Many aldehydes and ketones possess strong odors; for example, the odor of Formalin, a solution of formaldehyde (pictured above), the simplest aldehyde, may be familiar to biology students. Aromatic aldehydes possess a variety of odors; for instance, benzaldehyde, the simplest aromatic aldehyde, has an aroma of almonds. Vanillin, another aromatic aldehyde, is found in the seed pods of the vanilla plant, and acetone (pictured above), the simplest ketone, possesses the familiar odor of fingernail polish remover, in which it is used as the solvent.

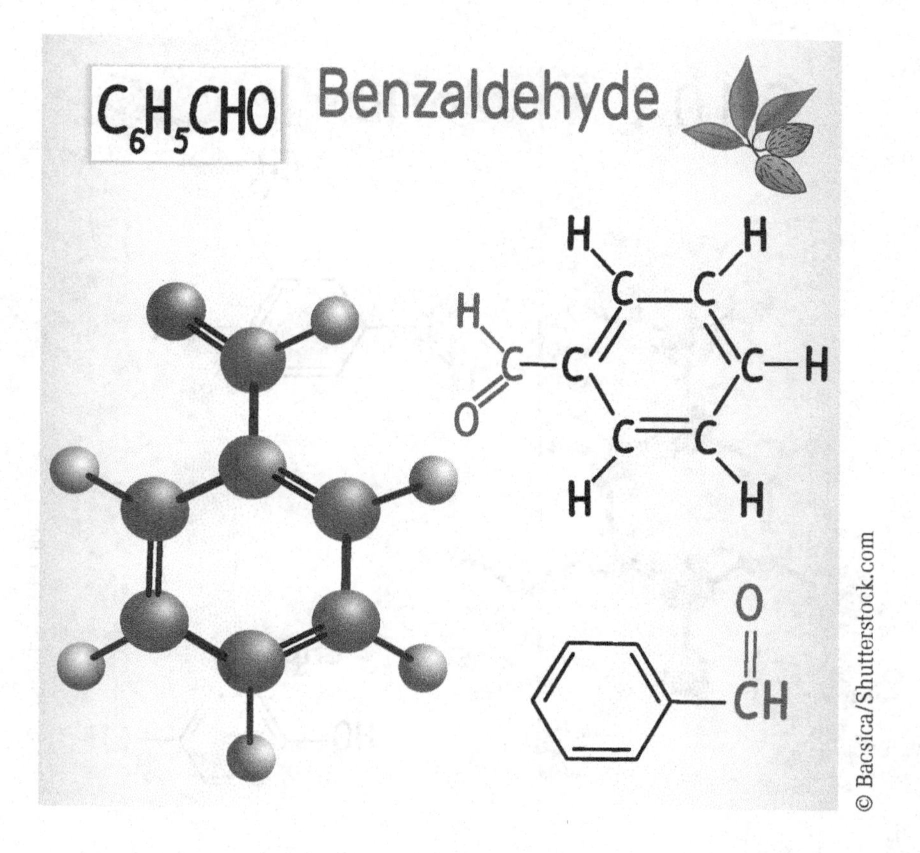

C. Solubility, Iodoform Test, and Benedict's Test

Ketones that contain a methyl group attached to the carbonyl carbon give a reaction with iodine (I_2) in a solution of NaOH, producing the yellow solid iodoform, CHI_3, possessing a strong medicinal odor often associated with hospitals, and used in gauze preparations as an antiseptic.

$CH_3\text{–}CO\text{–}CH_3 + 3I_3 + 4NaOH \rightarrow CH_3\text{–}COO^{-}Na^{+} + CHI_3 + 3NaI + 3H_2O$

Methyl ketone Iodine Iodoform

(Red) (Yellow)

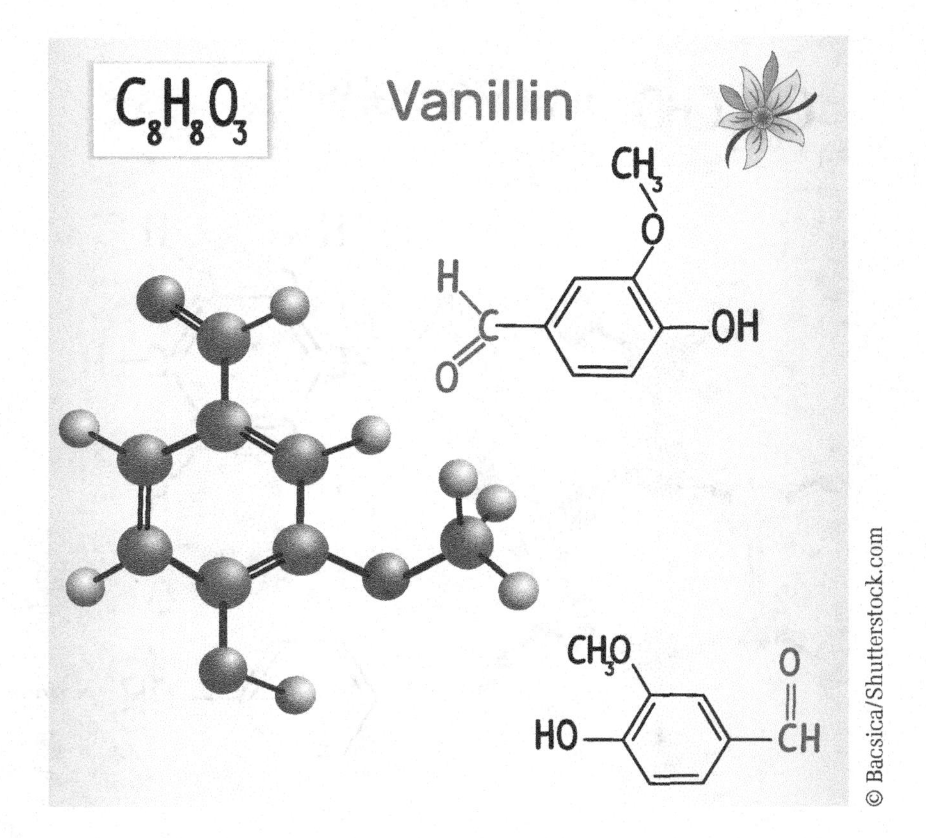

Oxidation of Aldehydes and Ketones

Aldehydes can be oxidized using Benedict's solution, which contains cupric ion, Cu^{2+}. Since ketones cannot undergo oxidation, this test can be used to distinguish aldehydes from ketones. In the oxidation reaction, the blue-green Cu^{2+} cupric ion is reduced to the Cu^{+} cuprous ion, which precipitates into reddish-orange Cu_2O.

$CH_3–COH + 2Cu^{2+} \rightarrow CH_3–COOH + Cu_2O(s)$

Aldehyde   Blue   Red-orange

$CH_3–CO–CH_3 + 2Cu^{2+} \rightarrow$ No reaction (stays blue)

Ketone   Blue

Identification of an Unknown

Using the results of these tests, an unknow substance can be identified as either an aldehyde or a ketone.

Experimental Procedures

A. Structures of Some Aldehydes and Ketones

Materials: Display of models: formaldehyde, acetaldehyde, propionaldehyde, acetone, butanone, and cyclohexanone.

B. Properties of Aldehydes and Ketones

Materials: 7 test tubes, 7 droppers, 5- or 10-mL graduated cylinder, acetone, formaldehyde, camphor, vanillin, cinnamaldehyde, propionaldehyde, and cyclohexanone, chemistry handbook, or Internet access.

1. Carefully detect the odor or samples of acetone, formaldehyde, camphor, vanillin, and cinnamaldehyde. Gently fan the fumes from above each test tube sample and record your observations.
2. Draw each of the condensed structural formulas; you may need to use a chemistry handbook or the Internet.
3. Identify each as an aldehyde or a ketone.

C. Solubility, Iodoform Test, and Benedict's Test

Solubility in Water

Materials: 5 test tubes, 5 droppers, stirring rod, 5- or 10-mL graduated cylinder, 10% NaOH, propionaldehyde, formaldehyde, acetone, cyclohexanone, and an unknown.

1. Place 2 mL of water in each of the 5 test tubes. Add 5 drops, respectively, of propionaldehyde, formaldehyde, acetone, cyclohexanone, and an unknown to each of the test tubes. Mix by gently stirring with stirring rod. If two separate

layers form or the mixture turns cloudy, the aldehyde or ketone is insoluble. Save test tubes for **Part C.2.**

Iodoform Test for Methyl Ketones

Materials: Test tubes from **Part C.1**, dropper, 10% NaOH, warm water bath, and iodine test reagent.

2. Using the test tube samples from **Part C.1**, add 10 drops of 10% NaOH to each. Warm the test tubes in a warm water bath to 50°C to 60°C. Add 20 drops of iodine test reagent. Mix by gently stirring with stirring rod. Look for the formation of a yellow precipitate and record your observations.
3. State whether a methyl ketone is present or not.

Benedict's Test

Materials: 5 test tubes, propionaldehyde, formaldehyde, acetone, cyclohexanone, unknown, Benedict's reagent, dropper, boiling water bath.

4. Place 10 drops, respectively, of propionaldehyde, formaldehyde, acetone, cyclohexanone, and unknown in each of the separate test tubes and label. Add 2 mL of Benedict's reagent to each test tube and mix. Place test tubes in a boiling water bath for 5 minutes. After 5 minutes, record the color of the samples. The appearance of the red-orange color of Cu_2O indicates that oxidation has occurred. Moderate amounts of Cu_2O will blend with the blue Cu^{2+} solution to give a green or rust color.
5. Identify the compounds that were oxidized.
6. Compare the results of the tests for the unknown with the tests performed with known aldehydes and ketones and identify the unknown.

Date ________________ Name ______________________________

Section ________________ Team ______________________________

Instructor ________________ ______________________________

Prelab Study Questions

1. Draw the condensed structural formula, if any, of the product from each of the following:

   a. $CH_3–CH_2–CH_2–COH + H_2 \xrightarrow{Pt}$

   b. $CH_3–CH_2–COH \xrightarrow{[O]}$

   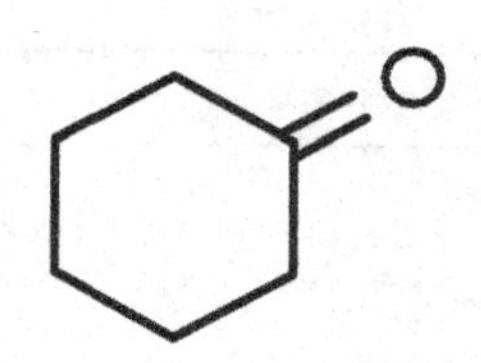

   c. cyclohexanone $+ H_2 \xrightarrow{Pt}$

   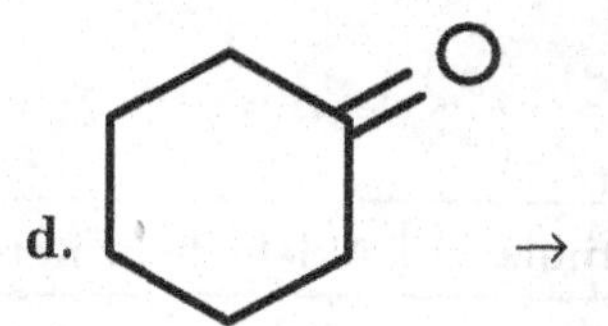

   d. cyclohexanone $\xrightarrow{[O]}$

2. Draw the condensed structural formula of each of the following:

   a. 2-Methylbutanal

   b. 3-Petanone

   c. 3-Bromopropanal

   d. 2-Methylcyclopentanone

Date ______________ Name ______________________________

Section ______________ Team ______________________________

Instructor ______________ ______________________________

Report Sheet for Lab 17: Aldehydes and Ketones

A. Structures of Some Aldehydes and Ketones

Formaldehyde	Acetaldehyde
IUPAC Name:	IUPAC Name:
Propionaldehyde	Acetone
IUPAC Name:	IUPAC Name:
Butanone	Cyclohexanone
Common Name:	

B. Properties of Aldehydes and Ketones

	Odor	Condensed Structural Formula	Aldehyde or Ketone?
Acetone			
Formaldehyde			
Camphor			

	Odor	Condensed Structural Formula	Aldehyde or Ketone?
Vanillin			
Cinnamaldehyde			

C. Solubility, Iodoform Test, and Benedict's Test

	Soluble?	Iodoform Test	Methyl Ketone?	Color	Oxidation? (Yes/No)
Propionaldehyde					
Formaldehyde					
Acetone					
Cyclohexanone					
Unknown					

LAB 18

Types of Carbohydrates

The major source of energy in our diets is carbohydrates. Foods such as potatoes, bread, pasta, and rice are high in carbohydrates. If more carbohydrates than are needed for energy are taken in, the excess is converted into fat, leading to weight gain. The carbohydrate family can be organized into three major classes: monosaccharides, disaccharides, and polysaccharides.

A. Monosaccharides

Monosaccharides contain C, H, and O atoms in empirical units of $(CH_2O)_n$. Most common monosaccharides contain six carbon atoms, known as hexoses, which have the general formula $C_6H_{12}O_6$. They contain many hydroxyl (–OH) groups along with a carbonyl group. Aldoses, as their name implies, are monosaccharides, which contain an aldehyde group, while ketoses contain a ketone group.

Monosaccharides		Sources
Glucose	$C_6H_{12}O_6$	Fruit juices, honey, corn syrup
Galactose	$C_6H_{12}O_6$	Lactose hydrolysis
Fructose	$C_6H_{12}O_6$	Fruit juices, honey, sucrose hydrolysis

The letters D and L refer to the chiral orientation of the hydroxyl (–OH) group on the carbon that is positioned one above the bottom carbon. In the isomers of glyceraldehyde, for instance, the position of the –OH is on the right in the D-isomer and on the left in the L-isomer.

The most important hexose monosaccharides are glucose, galactose, and fructose. D-glucose ($C_6H_{12}O_6$), the most common hexose, is also known as dextrose and is found in nature in fruits, vegetables, corn syrup, and honey; it is also the form of

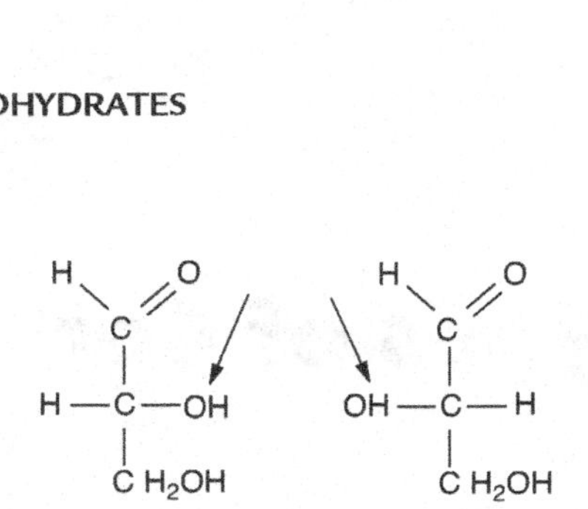

D-Glyceraldehyde L-Glyceraldehyde

Source: Bruce Kowiatek

glucose found in circulating blood, commonly referred to as blood sugar. D-glucose is a building block of disaccharides such as sucrose, lactose, and maltose, as well as polysaccharides such as amylose, cellulose, and glycogen. D-galactose ($C_6H_{12}O_6$), an aldohexose obtained from the disaccharide lactose, is found in milk and milk products. It is an important component of cellular membranes found in the brain and nervous system. The only difference between D-glucose and D-galactose lies in the arrangement of the –OH group on carbon 4. In contrast to D-glucose and D-galactose, D-fructose ($C_6H_{12}O_6$) is a ketohexose; its structure differs from D-glucose at carbons 1 and 2 by the location of the carbonyl group. The sweetest of the carbohydrates, D-fructose is nearly twice as sweet as sucrose (table sugar).

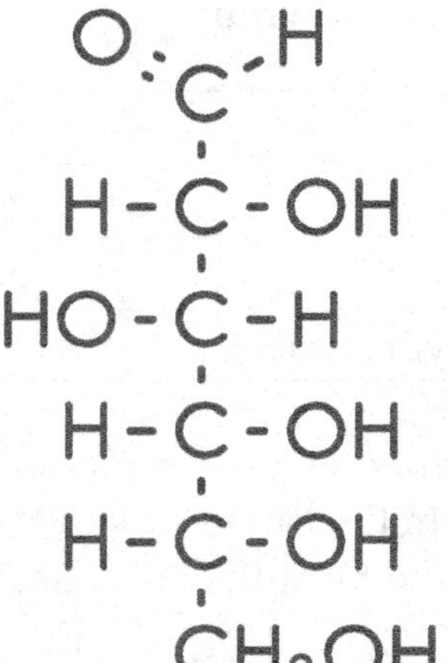

D-glucose

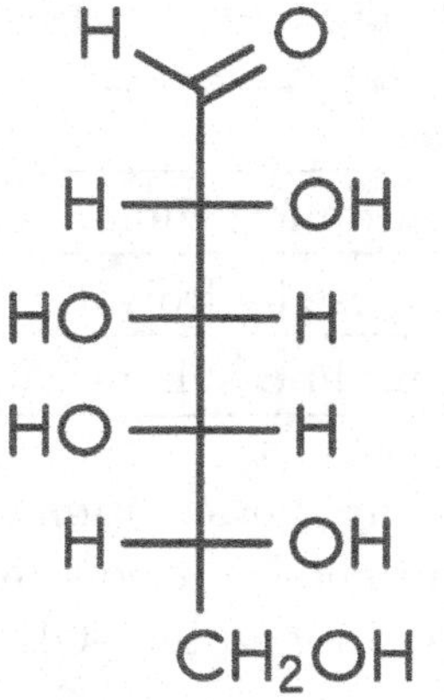

D-galactose

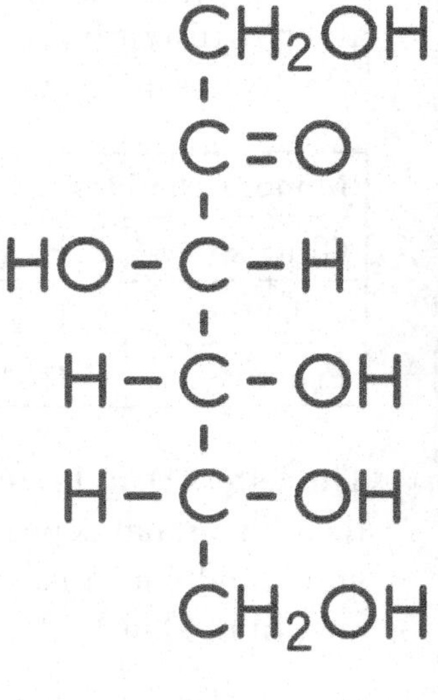

D-fructose

Haworth Structures

D-glucose exists in a ring structure most of the time, which forms when the –OH on carbon 5 bonds to carbon 1 in the carbonyl group. In the Haworth structure, the new hydroxyl group may be drawn below carbon 1 (the α form or anomer), or above carbon 1 (the β form or anomer).

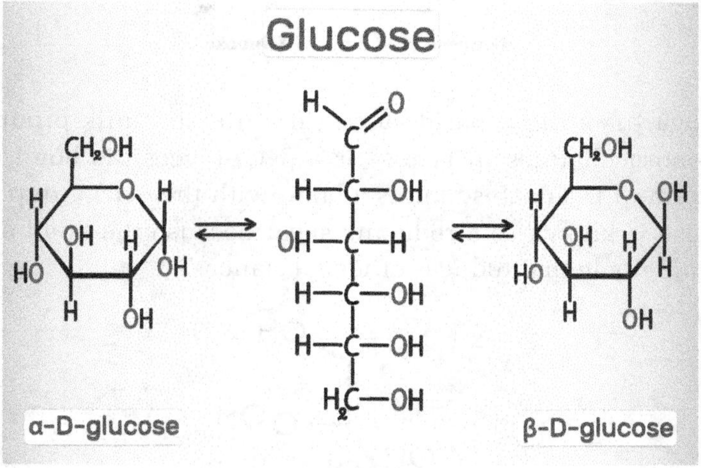

Glucose

α-D-glucose

β-D-glucose

B. Disaccharides

Two monosaccharides bonded together are known as a disaccharide. Common disaccharides include maltose, sucrose (table sugar), and lactose (milk sugar).

Disaccharides	Source	Monosaccharides
Maltose	Germinating grains, starch hydrolysis	D-glucose + D-glucose
Sucrose	Sugar cane, sugar beets	D-glucose + D-fructose
Lactose	Milk, yogurt, ice cream	D-glucose + D-galactose

In a disaccharide, two monosaccharides form a glycosidic bond with the loss of water in a condensation or dehydration reaction; for example, in maltose, two D-glucose units are linked by an α-1,4-glycosidic bond.

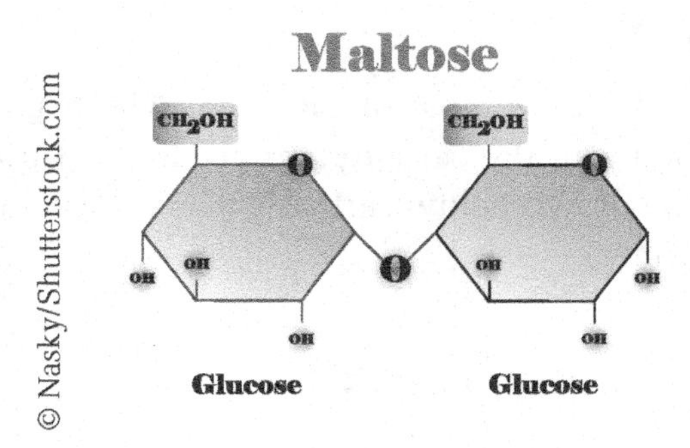

Lactose (milk sugar) is a disaccharide found in milk and milk products. The bond between the monosaccharides in lactose is a β-1,4-glycosidic bond since the –OH group of the β form of D-galactose forms a bond with the –OH group on carbon 4 of D-glucose. Lactose is known as a reducing sugar because the bond at carbon 1 can open to give an aldehyde that reduces other substances.

lactose

C. Polysaccharides

Long-chain polymers containing many thousands of monosaccharides, usually D-glucose, joined together by glycosidic bonds are known as polysaccharides. Starch, cellulose, and glycogen are three important polysaccharides; all contain D-glucose units, but differ in the type of glycosidic bonds and the amount of branching in the polymer.

Polysaccharides	Found in	Monosaccharides
Starch (amylose, amylopectin)	Rice, wheat, grains, cereals	D-glucose
Cellulose	Wood, plants, paper, cotton	D-glucose
Glycogen	Muscle, liver	D-glucose

Starch is an insoluble storage form of D-glucose found in rice, wheat, potatoes, beans, and cereals. It is composed of two kinds of polysaccharides, amylose and amylopectin. Making up approximately 20% of starch, amylose consists of α-D-glucose molecules connected by α-1,4-glycosidic bonds in a continuous chain. A typical polymer of amylose can contain anywhere from 250 to 4000 D-glucose units.

Making up the remaining 80% of starch, amylopectin is a branched-chain polysaccharide with α-1,4-glycosidic bonds connecting most of the D-glucose molecules; however, at around every 25 D-glucose units, there are branches of D-glucose molecules attached by α-1,6-glycosidic bonds between carbon 1 of the branch and carbon 6 in the main chain.

Cellulose comprises the major structural material of wood and plants. Cotton, for example, is almost pure cellulose, in which D-glucose molecules form a long, unbranched chain similar to amylose except that β-1,4-glycosidic bonds connect them. These β isomers align in parallel rows held in place by hydrogen bonds between the rows, giving a rigid structure for cell walls in wood and fiber, making cellulose more resistant to hydrolysis.

Similar to amylopectin but even more highly branched, glycogen possesses α-1, 6-glycosidic bonds roughly every 10 to 15 D-glucose units.

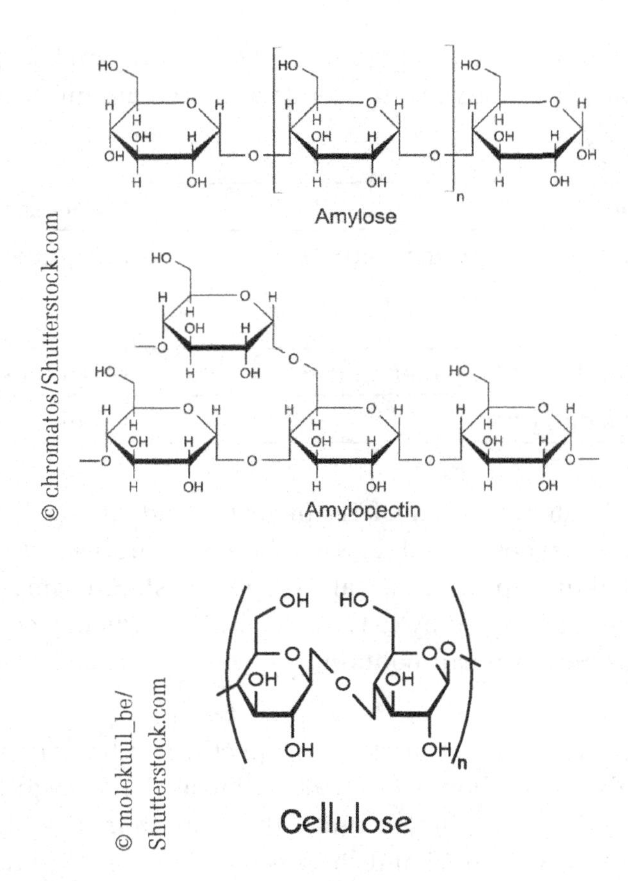

Experimental Procedures

A. Monosaccharides

Materials: Organic model kits or prepared models.

1. Make or observe models of L-glyceraldehyde and D-glyceraldehyde. Draw their Fischer projections.
2. Draw the Fischer projections for D-glucose, D-galactose, and D-fructose.
3. Draw the cyclic Haworth structures for the α and β anomers of D-glucose.
4. Draw the cyclic Haworth structures for the α anomers of D-galactose and D-fructose.

B. Disaccharides

1. Draw the Haworth structure for α-maltose.
2. Write an equation, using Haworth structures, for the hydrolysis of α-maltose by adding H_2O to the glycosidic bond.
3. Write an equation, using Haworth structures, for the formation of α-lactose from β-D-galactose and α-D-glucose.
4. Draw the Haworth structure of sucrose and circle the glycosidic bond.

C. Polysaccharides

1. Draw a portion of amylose, using Haworth structures, consisting of four units of α-D-glucose, and indicate the glycosidic bonds.
2. Draw a portion of cellulose, using Haworth structures, consisting of four units of β-D-glucose, and indicate the glycosidic bonds.

Date ________________ Name ________________________________

Section ________________ Team ________________________________

Instructor ________________ ________________________________

Prelab Study Questions

1. Name some sources of carbohydrates in your diet.

2. What does the D in D-glucose mean?

3. Name the bond that links monosaccharides in di- and polysaccharides.

4. Draw the Haworth structure of the following:

a. α-D-galactose

b. β-lactose

Date ________________ Name ________________________________

Section ________________ Team ________________________________

Instructor ________________ ________________________________

Report Sheet for Lab 18: Types of Carbohydrates

A. Monosaccharides

1.

L-Glyceraldehyde	D-Glyceraldehyde

2.

D-Glucose	D-Galactose	D-Fructose

3.

α-D-Glucose	β-D-Glucose

4.

α-D-Galactose	α-D-Fructose

Questions and Problems

1. Describe how the structure of D-glucose compares to the structure of D-galactose.

B. Disaccharides

1. α-maltose

2. Equation for the hydrolysis of α-maltose

3. Equation for the formation of α-lactose

4. Sucrose

Questions and Problems

1. What type of glycosidic bond is in maltose?

2. Explain why maltose has both α and β anomers.

C. Polysaccharides

1. Amylose

What is the difference between the structures of amylopectin and amylose?

2. Cellulose

Questions and Problems

1. What monosaccharide results from the complete hydrolysis of amylose?

2. What is the difference between the structures of amylose and cellulose?

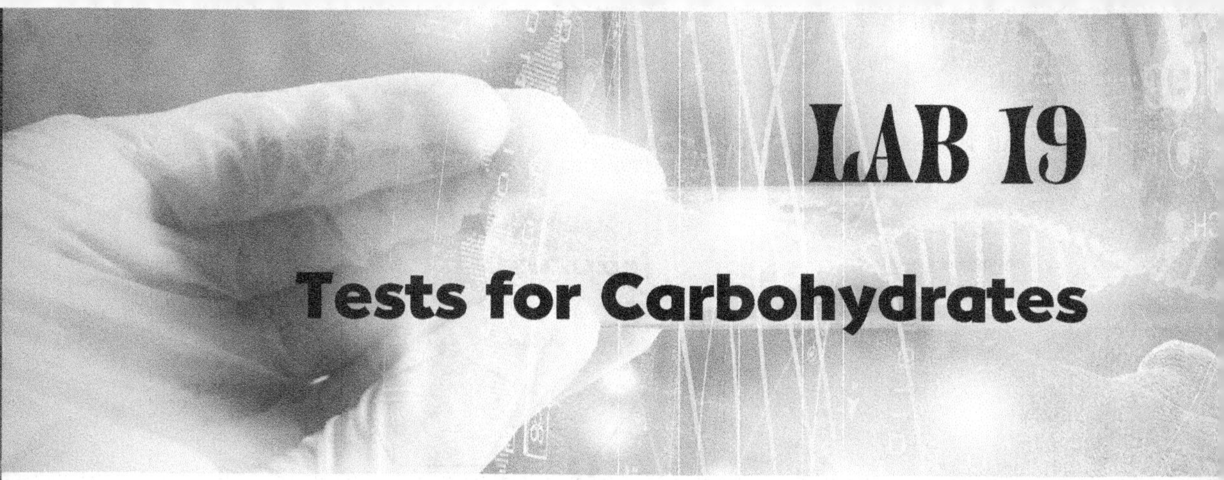

LAB 19

Tests for Carbohydrates

A. Benedict's Test for Reducing Sugars

All monosaccharides and most disaccharides can be oxidized to carboxylic acids. Upon opening of the cyclic structure, the aldehyde group is available for oxidation. Benedict's reagent contains reduced Cu^{2+} ion. All sugars that react with Benedict's reagent, therefore, are called *reducing sugars*. Ketoses can also act as reducing sugars since the ketone group on carbon 2 isomerizes to give an aldehyde group on carbon 1.

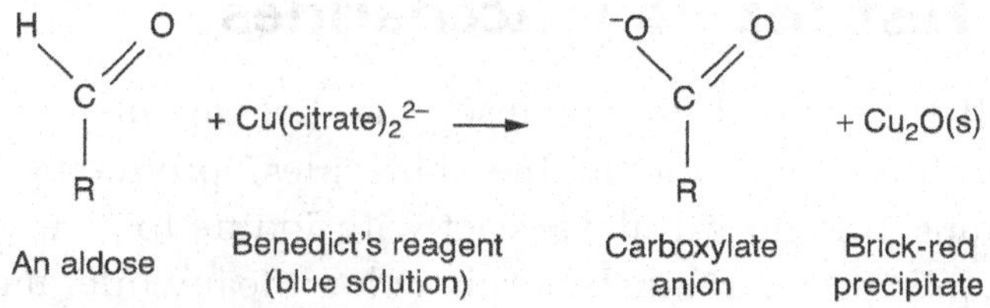

$$R\text{-}CHO + Cu(citrate)_2^{2-} \longrightarrow R\text{-}COO^- + Cu_2O(s)$$

An aldose; Benedict's reagent (blue solution); Carboxylate anion; Brick-red precipitate

Source: Bruce Kowiatek

When oxidation of a sugar occurs, Cu^{2+} is reduced to Cu^+, which forms a red precipitate of copper(I) oxide, $Cu_2O(s)$. The precipitate's color varies from green to gold to red, dependent upon the reducing sugar's concentration. If the reducing sugar's concentration is low, the sample takes on a gold or green color; if its concentration is high, the sample takes on a red color. Sucrose is not a reducing sugar because it cannot revert to the open-chain form that provides the aldehyde group necessary to reduce the copper(II) ion.

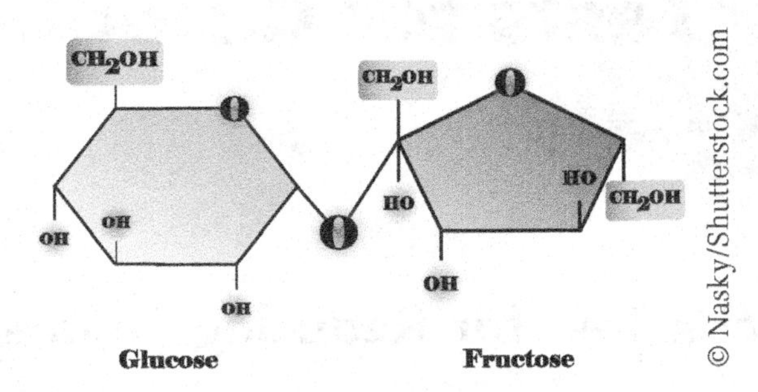

B. Seliwanoff's Test for Ketoses

Seliwanoff's test is used to distinguish between hexoses with a ketone group (ketohexoses) and hexoses that are aldehydes (aldohexoses). Ketohexoses form a deep red color rapidly, while aldohexoses develop a light pink color over a longer time period. The test is most sensitive for fructose, a ketohexose.

C. Iodine Test for Polysaccharides

When iodine (I_2) is added to amylose, the helical shape of the unbranched polysaccharide chain traps the iodine molecules, producing a deep blue-black complex. Amylopectin and cellulose react with iodine to give red-to-brown colors, while glycogen produces a reddish-purple color. Meanwhile, monosaccharides and disaccharides are too small to trap iodine molecules and form no dark colors with iodine.

D. Hydrolysis of Disaccharides and Polysaccharides

Disaccharides hydrolyze in the presence of an acid to give the individual constituent monosaccharides.

$$\text{Sucrose} + H_2O \xrightarrow{H^+} \text{Glucose} + \text{fructose}$$

In the lab, water and acid are used to hydrolyze starches, yielding smaller saccharides like maltose, and then the hydrolysis reaction eventually converts maltose to glucose molecules. In the body, enzymes in saliva and the pancreas carry out the hydrolysis reaction. Complete hydrolysis produces glucose, providing approximately 50% of our nutritional calories.

$$\text{Amylose, amylopectin} \xrightarrow[\text{Amylase}]{H^+ \text{ or}} \text{Dextrins} \xrightarrow[\text{Amylase}]{H^+ \text{ or}} \text{Maltose} \xrightarrow[\text{Maltase}]{H^+ \text{ or}} \text{Many glucose units}$$

Experimental Procedures

A. Benedict's Test for Reducing Sugars

Materials: 7 test tubes, 7 rubber test tube stoppers, 400-mL beaker for boiling water bath, 7 droppers, hot plate or propane torch, 5- or 10-mL graduated cylinder, Benedict's reagent, 2% solutions of glucose, fructose, lactose, sucrose, starch, and an unknown.

Place 10 drops of solution of water (reference), glucose, fructose, lactose, sucrose, starch, and an unknown in separate test tubes. Label each test tube.

Add 2 mL of Benedict's reagent to each sample, stopper, and mix by shaking. Remove stoppers and place all test tubes in a boiling water bath for 3 to 4 minutes.

Formation of a greenish to reddish-orange color indicates the presence of a reducing sugar; if the solution remains the same color as the Benedict's reagent, no oxidation reaction has occurred.

1. Record your observations.
2. Classify each as a reducing or nonreducing sugar.

B. Seliwanoff's Test for Ketones

Materials: 7 test tubes, 7 rubber test tube stoppers, 400-mL beaker for boiling water bath, 7 droppers, hot plate or propane torch, 5- or 10-mL graduated cylinder, Seliwanoff's reagent, 2% solutions of glucose, fructose, lactose, sucrose, starch, and an unknown.

Place 10 drops of solution of water (reference), glucose, fructose, lactose, sucrose, starch, and an unknown in separate test tubes. Label each test tube.

Add 2 mL of Seliwanoff's reagent to each sample, stopper, and mix by shaking. ***Use carefully, reagent contains concentrated hydrochloric acid (HCl).***
Place all test tubes in a boiling hot water bath and note the time.

1. After 1 minutes, observe the colors in the test tubes. Rapid formation of a deep red color indicates the presence of a ketohexose.
2. Classify each as either a ketohexose or aldohexose.

C. Iodine Test for Polysaccharides

Materials: Spot plate or 7 small test tubes, 7 droppers, iodine reagent, 2% solutions of glucose, fructose, lactose, sucrose, starch, and an unknown.

Using a spot plate or small test tubes, place five drops of water and each solution of glucose, fructose, lactose, sucrose, starch, and an unknown in the wells or test tubes.

Add one drop of iodine solution to each sample; a dark blue-black color indicates a positive test for amylose in the starch.

1. Record your observations.
2. Indicate whether amylose is present or not.

D. Hydrolysis of Disaccharides and Polysaccharides

Materials: 4 test tubes, 4 rubber test tube stoppers, 10-mL graduated cylinder, 400-mL beaker for boiling water bath, hot plate or propane torch, spot plate or watch glass, 10% HCl, 10% NaOH, red litmus paper, iodine reagent, Benedict's reagent, 2% starch and sucrose solutions.

Place 3 mL of 2% starch in 2 test tubes.

Place 3 mL of 2% sucrose in the other 2 test tubes.

To one sample each of sucrose and starch, add 20 drops (1 mL) of 10% HCl, stopper, and mix by shaking.

To the other 2 samples of sucrose and starch, add 20 drops (1 mL) of H_2O, stopper, and mix by shaking.

Label all test tubes and heat in a boiling water bath for 10 minutes.

Remove the test tubes from the boiling water bath and allow to cool.

To the samples containing HCl, add red litmus paper and enough 10% NaOH dropwise up to 20 drops or until 1 drop of the mixture changes the red litmus paper to blue, which indicates that the HCl has been neutralized.

Test each separate sample for hydrolysis using the iodine test and Benedict's test as described below:

Iodine Test

Place five drops of each solution on a spot plate or watch glass.
Add one drop of iodine reagent to each.

1. Record your observations.

Benedict's Test

Add 2 mL of Benedict's reagent to each of the samples.

Heat each sample in a boiling water bath for 3 to 4 minutes.

2. Record your observations.
3. Determine if hydrolysis has occurred in each or not.

Date ______________ Name ______________________________

Section ______________ Team ______________________________

Instructor ______________ ______________________________

Prelab Study Questions

1. What happens to the glucose or galactose when the Cu^{2+} in Benedict's reagent is reduced?

2. Would you expect fructose or glucose to form a red color rapidly with Seliwanoff's reagent? Explain your answer.

3. How can the iodine test be used to distinguish between amylose and glycogen?

Date ________________ Name ______________________________

Section ________________ Team ______________________________

Instructor ________________ ______________________________

Report Sheet for Lab 19: Tests for Carbohydrates

Test	Water	Glucose	Fructose	Lactose	Sucrose	Starch	Unknown
A. Benedict's Test for Reducing Sugars							
1. Observations							
2. Reducing/ nonreducing							
B. Seliwanoff's Test for Ketohexoses							
1. Colors after 1 minutes							
2. Ketohexose/ aldohexose							
C. Iodine Test for Polysaccharides							
1. Observations							
2. Amylose (yes/ no)							

Questions and Problems

1. From the results in **Part A**, list the carbohydrates that are reducing sugars.

2. From the results in **Part B**, which carbohydrates are ketohexose?

3. From the results in **Part C**, which carbohydrates give a blue-black color in the iodine test?

Identifying an Unknown Carbohydrate

	Results with Unknown	Carbohydrate Present
Benedict's **(A)**		
Seliwanoff's **(B)**		
Iodine **(C)**		

What carbohydrate(s) is/are in your unknown?

Questions and Problems

4. What carbohydrate(s) would give the following result?

   a. Produces a reddish-orange solid with Benedict's reagent and a red color with Seliwanoff's reagent in 1 minute.

b. Produces no color change with Benedict's or Seliwanoff's reagent but turns a blue-black color with iodine reagent.

D. Hydrolysis of Disaccharides and Polysaccharides

Results	Sucrose + H_2O	Sucrose + HCl	Starch + H_2O	Starch + HCl
Iodine test				
Benedict's test				
Hydrolysis (yes/no)				

Questions and Problems

5. How do the results of the Benedict's test indicate that hydrolysis of sucrose and starch occurred?

6. How do the results of the iodine test indicate that hydrolysis of starch occurred?

7. Indicate whether the following carbohydrates will give a positive (+) or a negative (–) result in each type of test listed below:

	Benedict's Test	Seliwanoff's Test	Iodine Test
Glucose			
Fructose			
Galactose			
Sucrose			
Lactose			
Maltose			
Amylose			

LAB 20

Carboxylic Acids and Esters

Salad dressings made from oil and vinegar taste sour due to the vinegar, which contains acetic acid (ethanoic acid). The sour taste associated with fruits like lemons is due to acids such as citric acid and ascorbic acid (Vitamin C). Many facial creams contain alpha hydroxy acids like glycolic acid. All these acids are carboxylic acids, which contain a carboxyl functional group: a carbonyl attached to a hydroxyl group. A dicarboxylic acid like malonic acid possesses two carboxylic acid functional groups. The carboxylic acid of benzene is known as benzoic acid.

A. Carboxylic Acids and Their Salts

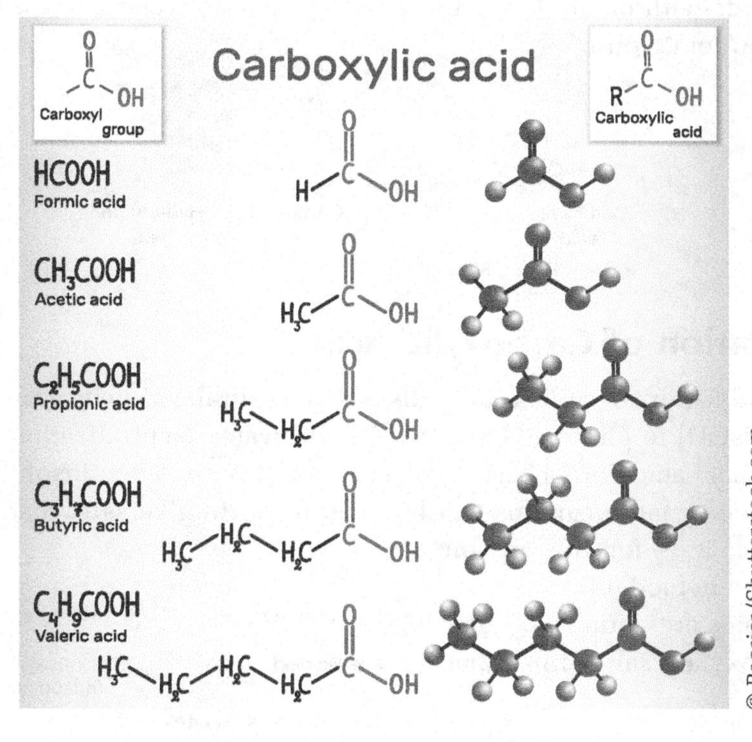

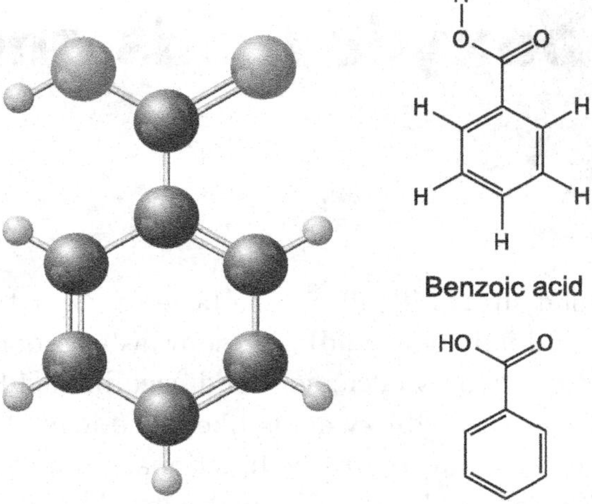

Benzoic acid

Ionization of Carboxylic Acids in Water

Carboxylic acids tend to be weak acids because the carboxylic acid group slightly ionizes in water to form a hydrogen ion (H^+) and the carboxylate ion shown in the below equation. Acids with one to four carbons are soluble in water due to the polarity of the carboxylic acid group.

$$\underset{\text{Carboxylic acid}}{RC(=O)-O-H} + \underset{\text{Water}}{H_2O} \rightleftharpoons \underset{\text{Carboxylate ion}}{RC(=O)-O^-} + \underset{\text{Hydronium ion}}{H_3O^+}$$

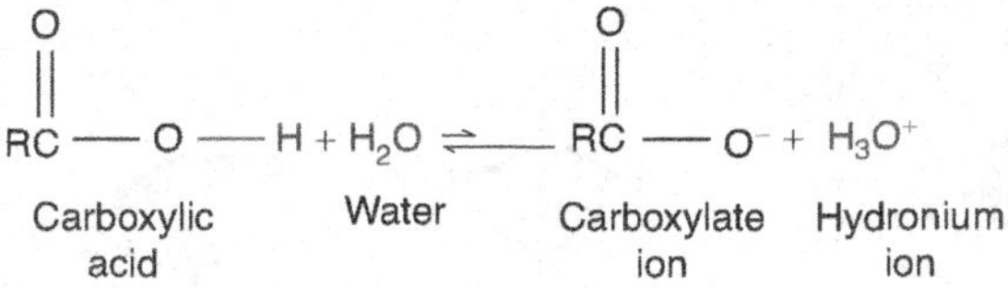

Source: Bruce Kowiatek

Neutralization of Carboxylic Acids

One important feature of carboxylic acids is their neutralization by a base like sodium hydroxide (NaOH) to form carboxylate salts and water. Neutralization once again is the reaction of an acid with a base to give a salt and water. Even insoluble carboxylic acids with five or more carbons can be neutralized to give salts that are generally water soluble. It is for this reason that acids used in food products and/or medications are found in their soluble carboxylate salt form rather than the acid.

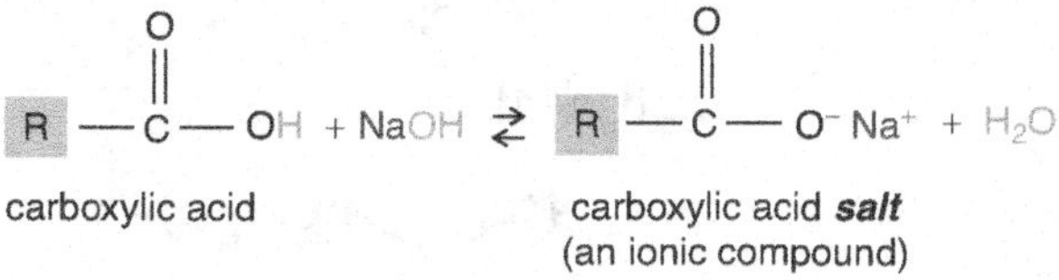

$$\underset{\text{carboxylic acid}}{R-C(=O)-OH} + NaOH \rightleftharpoons \underset{\text{carboxylic acid salt (an ionic compound)}}{R-C(=O)-O^-\,Na^+} + H_2O$$

Source: Bruce Kowiatek

B. Esters

While carboxylic acids can have tart or unpleasant odors, many esters have pleasant flavors and fragrant odors. Octyl acetate gives oranges their characteristic aroma and flavor; the flavor of pears is due to pentyl acetate, and isobutyl formate is responsible for the flavor and smell of raspberries.

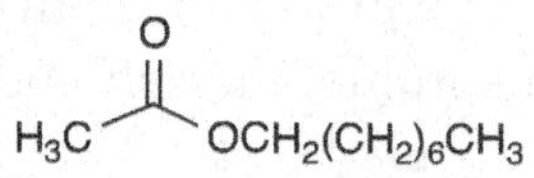

H_3C OCH$_2$(CH$_2$)$_6$CH$_3$

Octyl acetate (oranges)
Source: Bruce Kowiatek

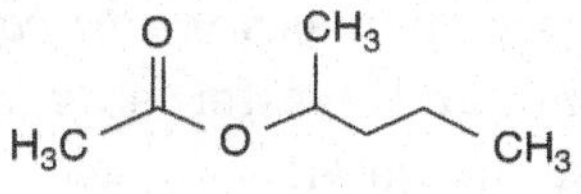

Pentyl acetate (pears)
Source: Bruce Kowiatek

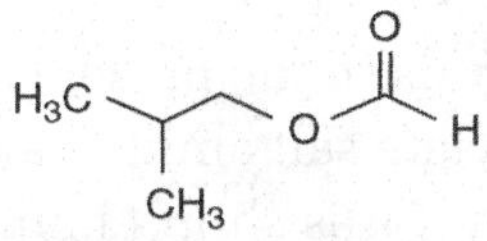

Isobutyl formate (raspberries)
Source: Bruce Kowiatek

Methyl salicylate is an ester of salicylic acid, which gives the flavor and odor of the oil of wintergreen that is used in candies and in ointments for sore muscles. When salicylic acid reacts with acetic anhydride, acetylsalicylic acid (ASA) is formed. Known more commonly as aspirin, it is used to reduce fever and inflammation.

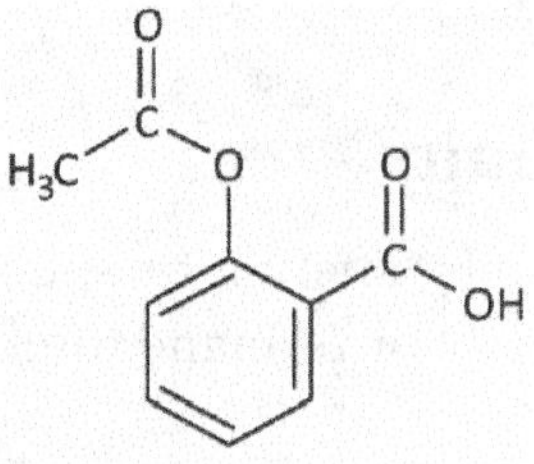

ACETYLSALICYLIC ACID

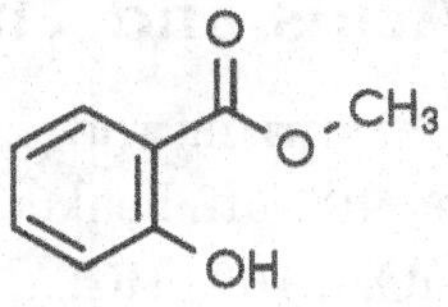

methyl salicylate

Esterification and Acid Hydrolysis

In the reaction known as *esterification*, the carboxylic acid group combines with the hydroxyl group of an alcohol. Taking place in the presence of an inorganic catalyst, the reaction produces an ester and water.

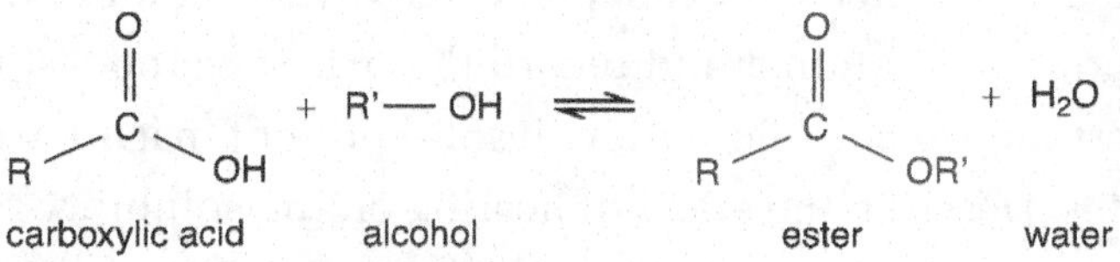

$$R\text{-}COOH + R'\text{-}OH \rightleftharpoons R\text{-}COOR' + H_2O$$

carboxylic acid, alcohol, ester, water

Source: Bruce Kowiatek

The reverse reaction, hydrolysis, takes place in acid or base. When an acid catalyst is used, the ester decomposes to yield the carboxylic acid and alcohol. The ester product is favored when an excess of acid or alcohol is used; hydrolysis is favored when excess water is present.

C. Saponification

Hydrolysis in the presence of a base is known as *saponification*. The products are the carboxylate salt and the alcohol. Although the ester is usually insoluble, the salt and alcohol, if it is short chain, are soluble in water.

$$\underset{\text{An ester}}{R-\overset{\overset{\large O}{\|}}{C}-OR} + \underset{\text{A base}}{NaOH\ (aq)} \longrightarrow \underset{\text{A carboxylate salt}}{R-\overset{\overset{\large O}{\|}}{C}-ONa\ (aq)} + \underset{\text{An alcohol}}{ROH}$$

Source: Bruce Kowiatek

Experimental Procedures

A. Carboxylic Acids and Their Salts

Materials: 2 test tubes, stirring rod, 2 droppers, spatula or spoonula, pH paper, red and blue litmus paper, 400-mL beaker, hot plate or propane torch, glacial acetic acid, benzoic acid, 10% NaOH, 10% HCl.

1. Write condensed structural formulas for acetic acid and benzoic acid
2. Place 2 mL of cold distilled water in 2 test tubes. Add 5 drops of acetic acid to one and a small amount of benzoic acid (enough to cover the tip of the spatula or spoonula) to the other. Tap the sides of the test tubes to mix or stir with a stirring rod. Record your observations. Save the test tubes for **Part 3**.
3. Test the pH of each carboxylic acid by dipping the stirring rod into the solution and touching it to a piece of pH paper. Compare the color of the paper with the color chart on the container and record the pH. Save the test tubes for **Part 4.**
4. Place the test tube of benzoic acid with solid present in a hot water bath, heating for 5 minutes. Describe the effect of heating on the solubility of the acid. Allow the test tube to cool and record your observations. Save the test tube for **Part 5.**

5. Add red litmus paper and enough drops of 10% NaOH to each test tube until the solution turns the red litmus paper blue. Record your observations and save the test tubes for **Part 6**.
6. Add blue litmus paper and enough drops of 10% HCl to each test tube until the solution turns the blue litmus paper red. Record your observations.

B. Esters

Materials: Models of acetic acid, methanol, and methyl acetate. Five test tubes, test tube rack, hot plate or propane torch, 400-mL beaker, stirring rod, spatula or spoonula, methanol, 1-pentanol, 1-octanol, benzyl alcohol, 1-propanol, H_3PO_4, salicylic acid.

Caution: Use care in dispensing glacial acetic acid, as it may cause burns and blisters on the skin.

1. Observe the models of acetic acid, methanol, and methyl acetate. Write the equation for the formation of methyl acetate.
2. Prepare the following mixtures by placing 3 mL of each alcohol in a test tube and labeling it with the corresponding letter.

Mixture	Alcohol	Carboxylic Acid
A	Methanol	Salicylic acid
B	1-Pentanol	Glacial acetic acid
C	1-Octanol	Glacial acetic acid
D	Benzyl alcohol	Glacial acetic acid

3. Add either 2 mL of liquid or the amount of solid that covers the tip of a spatula or spoonula of each carboxylic acid to the alcohol. With the test tubes pointed away from you and the other students, ***cautiously*** add 15 drops of concentrated phosphoric acid (H_3PO_4). Gently stir and then place the test tubes in a hot water bath for 15 minutes. Remove the test tubes and ***cautiously*** fan the vapors toward you. Record the odors detected.
4. Write the condensed structural formulas and names of the esters produced.

Dispose of the ester products in the Organic Waste bottle.

C. Saponification

Materials: Test tube, test tube holder, 2 droppers, stirring rod, hot plate or propane torch, 250- or 400-mL beaker, methyl salicylate, 10% NaOH, 10% HCl, blue litmus paper.

1. Draw the condensed structural formula of methyl salicylate.
2. Place 3 mL of distilled water in a test tube, add 5 drops of methyl salicylate, and record the appearance and odor of the ester.
3. Add 1 mL (20 drops) of 10% NaOH to the solution. Two layers should form in the test tube. Place the test tube in a hot water bath for 30 minutes or until the top layer of the ester disappears. Record any changes in the odor of the ester. Remove the test tube and cool in cold water.
4. Write the equation for the saponification reaction.
5. Upon cooling of the solution, add blue litmus paper and enough drops of 10% HCl until a drop of the solution turns the blue litmus paper red, up to 20 drops (1 mL). Record your observations.
6. Draw the condensed structural formula of the solid that forms.

Date ______________ Name ______________________________

Section ______________ Team ______________________________

Instructor ______________ ______________________________

Prelab Study Questions

1. Write the condensed structural formula of the organic product for propanoic acid reacting with each of the following:

   a. NaOH

   b. CH_3–OH

   c. H_2O

2. Write the condensed structural formula of the organic products for ethyl ethanoate when it reacts with each of the following:

   a. NaOH

   b. H_2O and HCl

Date ______________ Name ______________________

Section ______________ Team ______________________

Instructor ______________ ______________________

Report Sheet for Lab 20: Carboxylic Acids and Esters

A. Carboxylic Acids and Their Salts

	Acetic Acid	Benzoic Acid
Condensed structural formulas		
Solubility in cold water		
pH		
Solubility in hot water		
NaOH		
HCl		

B. Esters

Use condensed structural formulas to write the equation for the formation of methyl acetate.

Alcohol and Carboxylic Acid	Odor of Ester	Condensed Structural Formula and Name of Ester
Methanol and salicylic acid		
1-Pentanol and acetic acid		
1-Octanol and acetic acid		
Benzyl alcohol and acetic acid		
1-Propanol and acetic acid		

C. Saponification

Condensed structural formula
Appearance and odor
Changes in appearance and odor
Equation for saponification
Observations
Condensed structural formula of the compound formed by adding HCl

LAB 21

Lipids

Lipids comprise a family of biomolecules having the common property of being soluble in organic solvents (lipophilic) but not in water (hydrophobic). The term "lipid" derives from the Greek word *lipos*, meaning "fat" or "hard." Within the lipid family, specific structures distinguish different types of lipids. Lipids like waxes, fats, oils, and triacylphospholipids are esters, which can be hydrolyzed to yield fatty acids and the sugar alcohol *glycerol*. Steroids, on the other hand, have a completely different structure, do not contain fatty acids, and cannot be hydrolyzed. Steroids are characterized by a *steroid nucleus* composed of four fused carbon rings.

A. Physical Properties of Lipids and Fatty Acids

Lipids comprise a family of compounds grouped together by their similarities in solubility rather than their structures. As a group, lipids are generally more soluble in nonpolar solvents like ether, chloroform, and benzene than in water. Important types of lipids include fats and oils, glycerophospholipids, and steroids. Compounds classified as lipids include the fat-soluble vitamins, A. D, E, and K, cholesterol, hormones, portions of cell membranes, and vegetable oils. Table 21.1 lists the classes of lipids:

Table 21.1 Classes of Lipid Molecules

Lipids	Components
Waxes	Fatty acids and long-chain alcohol
Fats and oils (triacylglycerols)	Fatty acids and glycerol
Glycerophospholipids	Fatty acids, glycerol, phosphate, amino alcohol
Sphingolipids	Fatty acids, sphingosine, phosphate, amino alcohol
Steroids	A fused ring of structure of three cyclohexanes and a cyclopentane

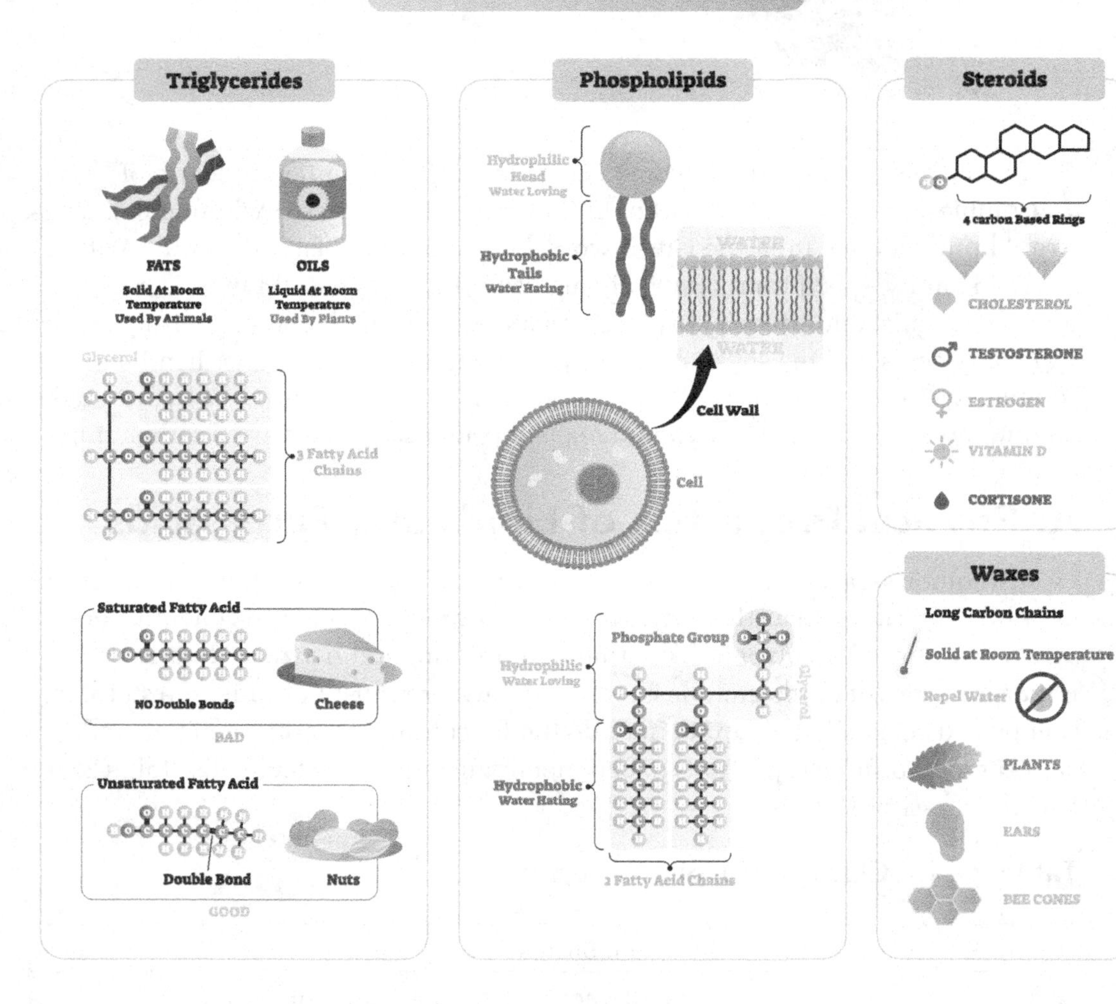

The structural formulas of three typical lipids are shown below:

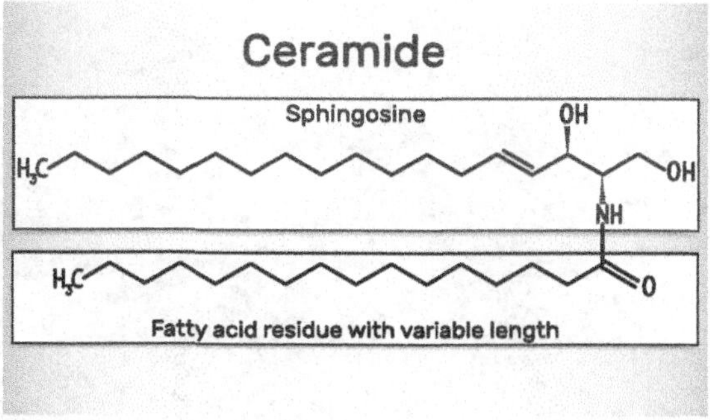

Ceramide
Sphingosine
OH
H_3C
OH
NH
H_3C
O
Fatty acid residue with variable length

Ceramide, a wax

H_3C
CH_3
O
O
O
O
O
O
CH_3
glyceryl tristearate

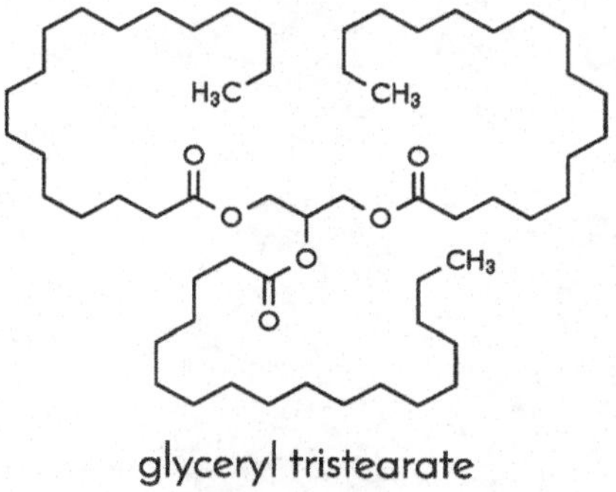

Glyceryl tristearate, a triacylglycerol

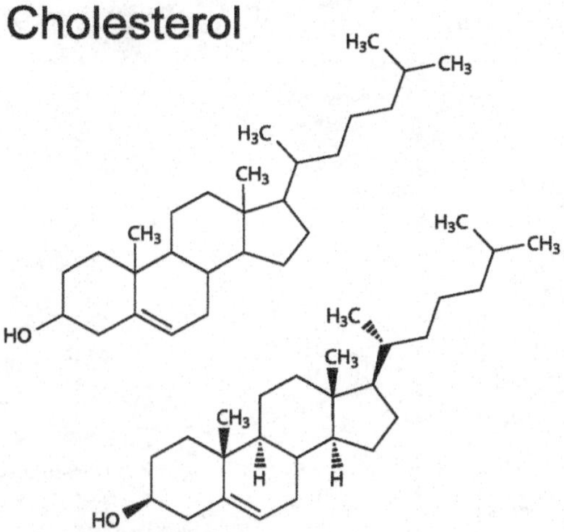

Cholesterol
H_3C
CH_3
H_3C
CH_3
CH_3
HO
H_3C
CH_3
H_3C
CH_3
CH_3
H
H
HO

Cholesterol, a steroid

B. Triacylglycerols

Fatty acids are long-chain carboxylic acids, typically 12 to 18 carbons long.

The triacylglycerols, or triglycerides, are esters of glycerol and fatty acids. When a fatty acid contains one or more carbon–carbon double bonds, the triacylglycerol is referred to as an unsaturated fat; when the fatty acid consists of only carbon–carbon single bonds, the triacylglycerol is a saturated fat.

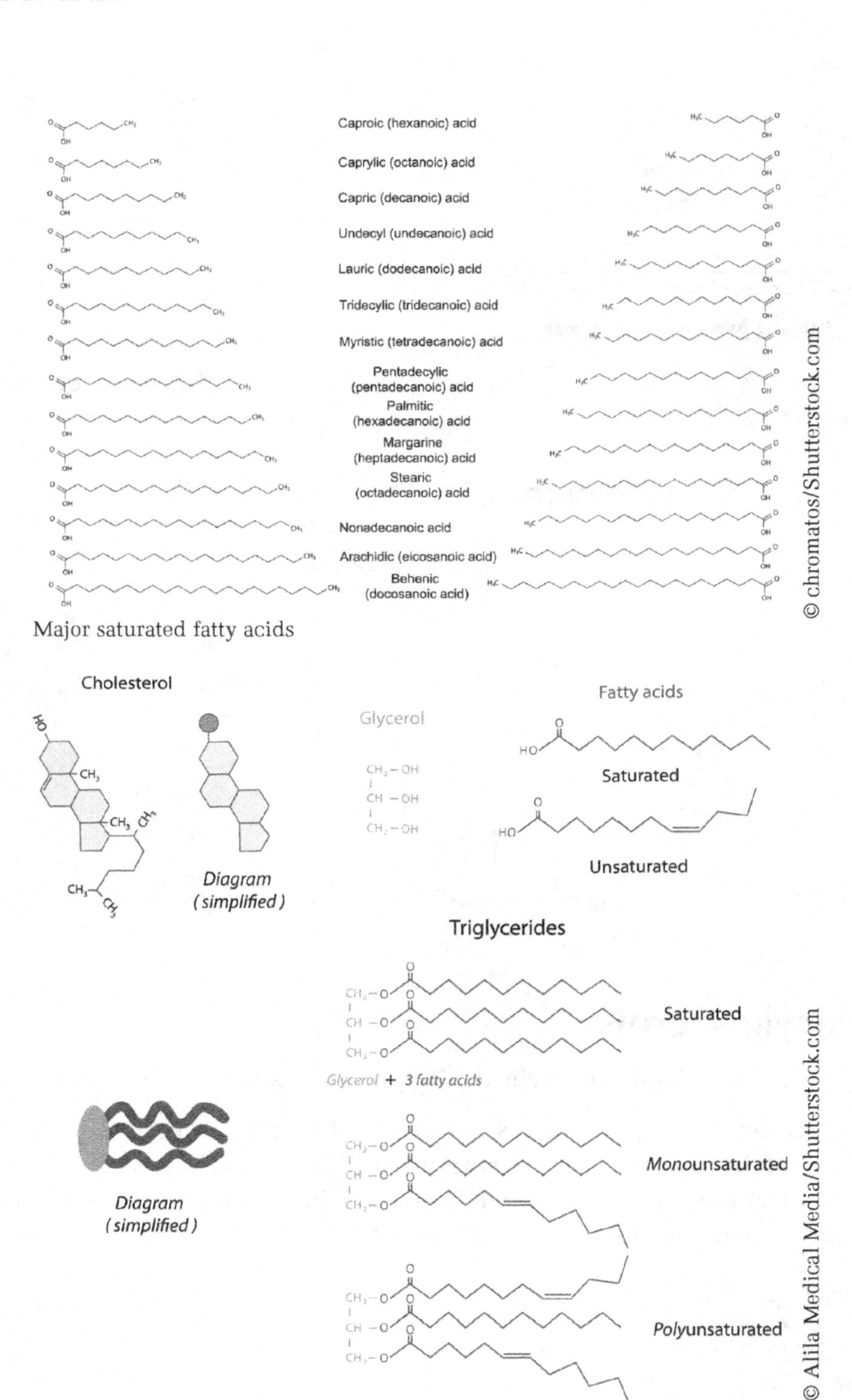

Caproic (hexanoic) acid
Caprylic (octanoic) acid
Capric (decanoic) acid
Undecyl (undecanoic) acid
Lauric (dodecanoic) acid
Tridecylic (tridecanoic) acid
Myristic (tetradecanoic) acid
Pentadecylic (pentadecanoic) acid
Palmitic (hexadecanoic) acid
Margarine (heptadecanoic) acid
Stearic (octadecanoic) acid
Nonadecanoic acid
Arachidic (eicosanoic acid)
Behenic (docosanoic acid)
Major saturated fatty acids
Cholesterol
Diagram (simplified)
Glycerol
Fatty acids
Saturated
Unsaturated
Triglycerides
Saturated
Glycerol + 3 fatty acids
Monounsaturated
Polyunsaturated
Diagram (simplified)

C. Bromine Test for Unsaturation

The presence of unsaturation in a fatty acid or triacylglycerol may be detected by the bromine test, which we've used earlier to detect double bonds in alkenes. If the orange color of the bromine solution fades quickly, an addition reaction has taken place and the oil or fat is unsaturated.

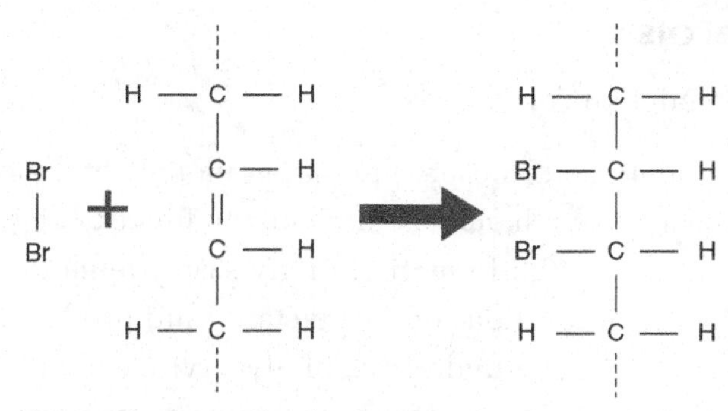

Source: Bruce Kowiatek

Experimental Procedures

A. Physical Properties of Lipids and Fatty Acids

Materials: 12 test tubes, 12 rubber test tube stoppers, stearic acid, oleic acid, safflower oil, lecithin, cholesterol, spatulas or spoonulas, CH_2Cl_2 (methylene chloride or dichloromethane).

1. To 6 separate test tubes, add 5 drops or the amount of solid lipid held on the tip of the spatula or spoonula of stearic acid, oleic acid, olive oil, safflower oil, lecithin, and cholesterol. Classify each as a triacylglycerol, fatty acid, steroid, or glycerophospholipid.
2. Describe the appearance of each.
3. Describe the odor of each.
4. Add 2 mL of distilled water to each test tube, stopper, and shake. Record the solubility or insolubility of the lipids in the polar solvent, water.

5. To 6 separate test tubes, add 5 drops or the amount of solid lipid held on the tip of the spatula or spoonula of stearic acid, oleic acid, olive oil, safflower oil, lecithin, and cholesterol. Add 1 mL (20 drops) of methylene chloride (dichloromethane), CH_2Cl_2 to each sample. Record the solubility or insolubility of the lipids in the nonpolar solvent.

B. Triacylglycerols

Materials: Organic model kits or models.

1. Use organic model kits or observe prepared models of a molecule of glycerol and three molecules of ethanoic acid. State the functional groups of each.
2. Write the equation for the formation of glyceryl triethanoate by drawing the condensed structural formulas of the reactants and products.
3. Write the equation for the hydrolysis of glyceryl triethanoate by drawing the condensed structural formulas of the reactants and products.

C. Bromine Test for Unsaturation

Materials: Samples from **Part A.5**, 1% Br_2 in methylene chloride

1. To the samples of stearic acid, oleic acid, safflower oil, and olive oil from **Part A.5**, add 1% bromine solution dropwise until either a permanent red-orange color is obtained or until 20 drops (1 mL) have been added.

   Caution: Avoid contact with bromine solution and breathing its fumes; it can cause painful burns.

   Determine if the red-orange color persists or fades rapidly and record your results.
2. Indicate whether each sample contains a saturated or unsaturated lipid or not.

Date ______________ Name ______________________________

Section ______________ Team ______________________________

Instructor ______________ ______________________________

Prelab Study Questions

1. What is the functional group in a triacylglycerol (triglyceride)?

2. Draw the skeletal formula for linolenic acid. Why is it an unsaturated fatty acid?

3. What type of solvent is required to remove a spot of oil and why?

4. Write the equation for the esterification of glycerol and three palmitic acids.

Date ________________ Name ________________________________

Section ________________ Team ________________________________

Instructor ________________ ________________________________

Report Sheet for Lab 21: Lipids

A. Physical Properties of Lipids and Fatty Acids

Lipid	Type	Appearance	Odor	Soluble in Water (Yes/No)	Soluble in CH_2Cl_2 (Yes/No)
Stearic acid					
Oleic acid					
Olive oil					
Safflower oil					
Lecithin					
Cholesterol					

Questions and Problems

1. Why are the compounds in **Part A** classified as lipid?

B. Triacylglycerols

1. Functional group of glycerol ____________________________

Functional group of ethanoic acid ____________________________

2. Equation for the esterification of glycerol and three ethanoic acids:

3. Equation for the hydrolysis of glyceryl triethanoate:

C. Bromine Test for Unsaturation

	Color Persists/Fades Rapidly	Saturated/Unsaturated
	Fatty Acids	
Stearic acid		
Oleic acid		
	Triacylglycerols	
Safflower oil		
Olive oil		

Questions and Problems

2. **a.** Draw the condensed structural formulas of stearic acid and oleic acid.

Stearic acid

Oleic acid

b. From the results of **Part C**, which is the more saturated fatty acid, stearic acid or oleic acid? Explain your reasoning.

c. From the results of **Part C**, were safflower oil and olive oil saturated or unsaturated? Explain your reasoning.

LAB 22

Amines and Amides

Amines are considered derivatives of ammonia in which one or more hydrogen atoms are replaced with alkyl or aromatic groups. The number of alkyl groups that are attached to the nitrogen atom determines its classification as either a primary (1°) (methylamine), secondary (2°) (dimethylamine), or tertiary (3°) (trimethylamine) amine, and aromatic amines use the name *aniline*

.

CH_3NH_2

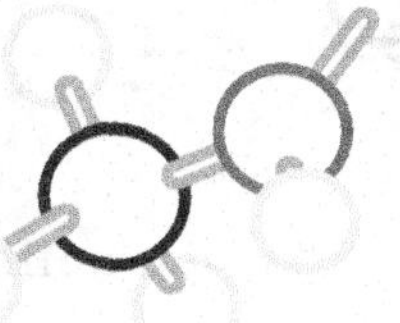

Methylamine

NH_3

Ammonia

$(CH_3)_2NH$

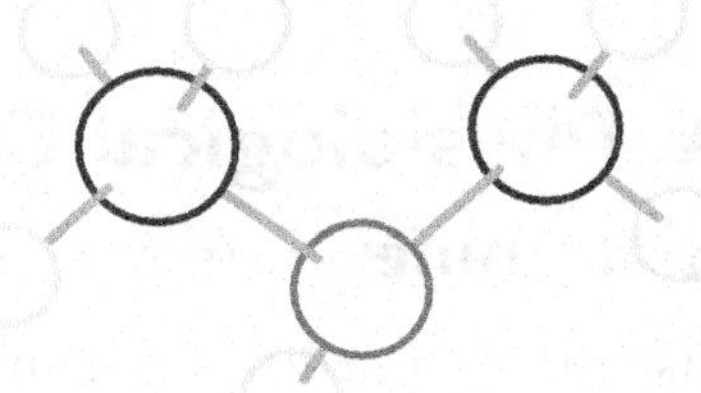

Dimethylamine

$N(CH_3)_3$

Trimethylamine

$C_6H_5NH_2$

Aniline

Amines can often be found as parts of compounds that are physiologically active (histamine) or used in medications (methamphetamine).

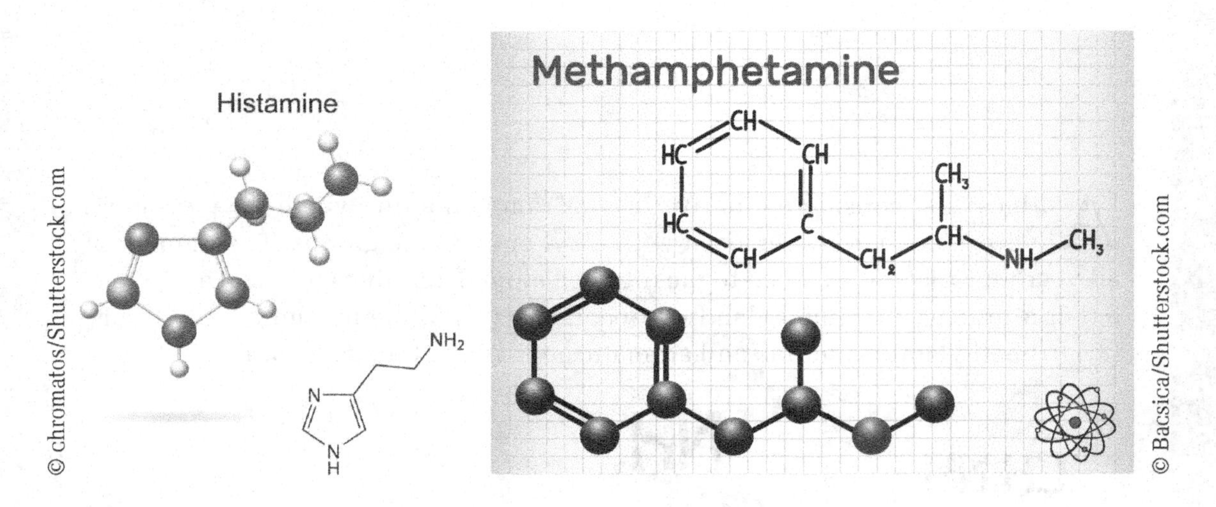

A. Physiological Effects of Some Amines

Histamine

Histamine is synthesized in nerve cells of the hypothalamus from the amino acid l-histidine when the histidine decarboxylase enzyme converts its carboxylate group to CO_2. It is produced by the immune system in response to allergens, pathogens, or injury. When it combines with H_1 histamine receptors, it causes allergic reactions that may include inflammation, watery eyes, itchy skin, and hay fever. Corollary to this, histamine may also cause smooth muscle contractions such as closing of the trachea in people allergic to shellfish. It is also stored and released in stomach cells where it stimulates acid production when it binds with H_2 receptors there. Antihistamines like Benadryl, Claritin, and Zyrtec block H_1 histamine receptors, stopping allergic reactions. So-called antacids such as Zantac, Pepcid, and Axid block H_2 histamine receptors, reducing stomach acid.

Histidine

Histidine decarboxylase → Histamine

Serotonin

Serotonin (5-hydroxytryptamine) enables us to relax, sleep deeply and peacefully, think rationally, and gives us a feeling of calmness and well-being. It is synthesized from the amino acid l-tryptophan, which crosses the blood–brain barrier. A diet containing foods like eggs, fish, cheese, turkey, chicken, and beef, which all have high levels of l-tryptophan and will increase serotonin levels. Foods with low levels of l-tryptophan, such as whole wheat, will lower serotonin levels. Psychedelic drugs like Lysergic acid diethylamide (LSD) and mescaline stimulate the action of serotonin at its receptors.

Low levels of serotonin in the brain are associated with depression, anxiety disorders, obsessive–compulsive disorder, and eating disorders. Many antidepressant drugs like Prozac (fluoxetine) and Paxil (paroxetine) are selective serotonin reuptake inhibitors (SSRIs). When reuptake of serotonin by brain neurons is slowed, it remains at the receptors longer, continuing its action, with the net effect being as if additional quantities of serotonin were taken.

Serotonin

Solubility of Amines

Ammonia and amines with one to four carbons act like weak bases in water because the unshared pair of electrons on the nitrogen atom attracts protons. The products are an ammonium ion or alkyl ammonium ion and a hydroxide ion.

$$NH_3 + H_2O \rightarrow NH_4^+ + OH^-$$

Ammonia — Ammonium ion — Hydroxide ion

$$CH_3\text{–}NH_2 + H_2O \rightarrow CH_3\text{–}NH_3^+ + OH^-$$

Methylamine — Methylammonium ion — Hydroxide ion

B. Neutralization of Amines with Acid

Since amines are basic, they react with acids to form an amine salt. These amine salts tend to be much more water soluble than their corresponding amines.

$$CH_3\text{–}NH_2 + HCl \rightarrow CH_3\text{–}NH_3^+\,Cl^-$$

Methylamine — Methylammonium chloride (Amine salt)

C. Amides

When a carboxylic acid reacts with ammonia or an amine, the product formed is an amide; therefore, the functional group is known as an amide group. In a reaction called *amidation*, an amide forms when a carboxylic acid is heated with ammonia or an alkyl or aromatic amine.

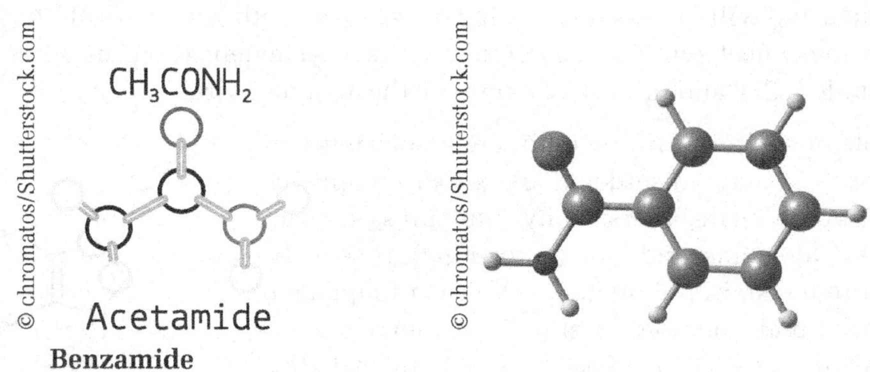

Benzamide

D. Hydrolysis of an Amide

When an amide is hydrolyzed, the amide bond is breaks and the carboxylic acid and amine are separated. Hydrolysis takes place in either an acid or a base. Acid hydrolysis produces the carboxylic acid and the ammonium salt. Base hydrolysis produces the carboxylic acid salt and ammonia. The odor or ammonia and the reaction of ammonia with litmus paper can be used to detect the hydrolysis reaction.

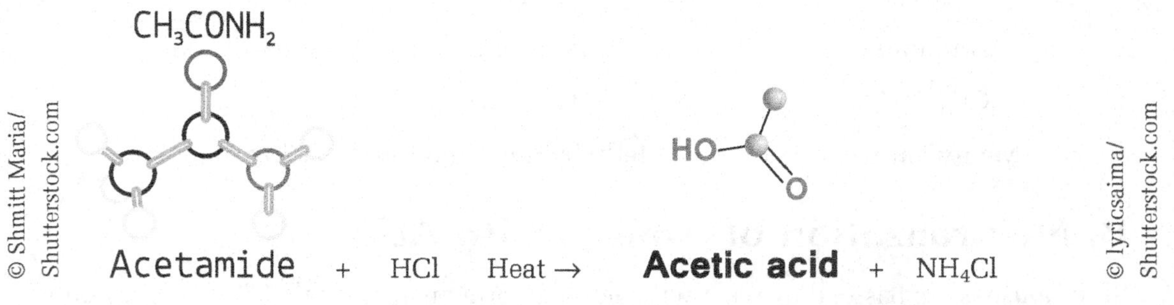

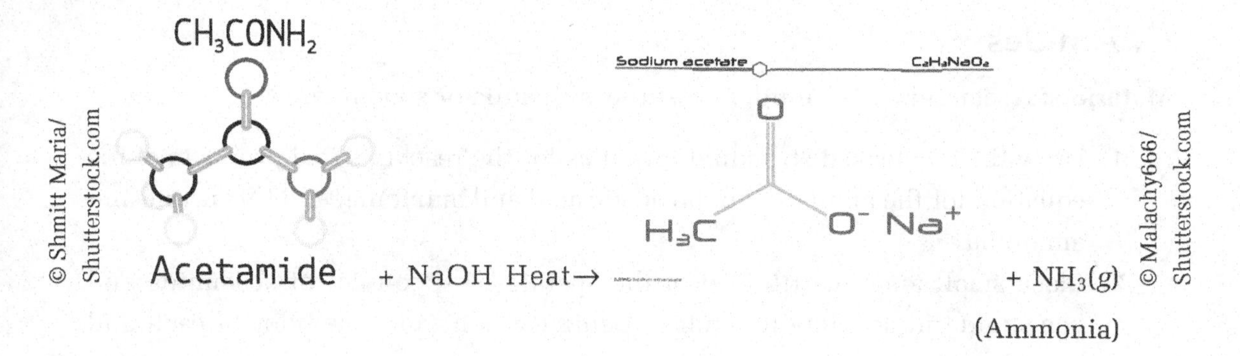

Experimental Procedures

A. Structure, Classification, and Solubility of Amines

Materials: Aniline, *N*-methylaniline, triethylamine, 3 test tubes, test tube rack, stirring rod, pH paper.

1. Draw the condensed structural formulas of aniline, *N*-methylaniline, and triethylamine.
2. State the classification (1°, 2°, or 3°) of each.
3. To 3 separate test tubes, add 5 drops each of aniline, *N*-methylaniline, and triethylamine, respectively. Cautiously fan the odor of each, remembering to hold a fresh breath of air while fanning the vapors toward you. Record the odors.
4. Add 2 mL of water to each test tube and stir. Clean and dry the stirring rod between test tubes and describe the solubility of each amine in water.
5. Determine the pH of each solution by dipping the stirring rod in each solution and then touching it to the pH paper. Record the pH of each.
   Save these test tubes and samples for Part B.

B. Neutralization of Amines with Acid

Materials: Test tubes from **Part A**, blue litmus paper, 10% HCl.

1. Using the amine solutions from **Part A**, add 10% HCl dropwise to each until the blue litmus paper turns red. Record any changes in the solubility of each amine.
2. Record any changes in odor.
3. Draw the condensed structural formulas for the reactants and products for the neutralization equations for aniline, *N*-methylaniline, and triethylamine with HCl.

C. Amides

Materials: Acetamide, benzamide, 2 test tubes, spatula or spoonula.

1. Draw the condensed structural formulas for the reactants and products of the equation for the amidation of (a) acetic acid and ammonia, and (b) aniline and ammonia.
2. Place small amounts (the tip of the spatula or spoonula) of acetamide and benzamide in separate test tubes. Using caution, note the odor of each and record.
3. Add 2 mL of water to each test tube and record the solubility of each amide in water.
   *Save these test tubes for **Part D.***

D. Hydrolysis of an Amide

Materials: Acetamide, benzamide, 2 test tubes, spatula or spoonula, 10% HCl, 10% NaOH, 250-mL beaker for boiling water bath, hot plate, red litmus paper, 10-mL graduated cylinder.

1. Draw the condensed structural formulas for the reactants and products of the equation for the hydrolysis of (a) acetamide and (b) benzamide using HCl.
2. Using the test tubes from **Part C.3**, add 2 mL of 10% HCl to each and place all of them in a boiling water bath, heating gently for 5 minutes. Using caution, note any odor coming from each mixture and record your observations.
3. Place small amounts (the tip of a spatula or spoonula) of acetamide and benzamide in separate test tubes. Add 2 mL of 10% NaOH to each, wet a piece of red litmus paper with distilled water, placing it over the mouth of each test tube, and place both in a boiling water bath, heating gently for 5 minutes. Record any changes in the color of the litmus paper.
4. Using caution, note any odor coming from each mixture and record your observations.
5. Draw the condensed structural formulas for the reactants and products of the equation for the hydrolysis of (a) acetamide and (b) benzamide using NaOH.

Date ______________ Name ______________________________

Section ______________ Team ______________________________

Instructor ______________ ______________________________

Prelab Study Questions

1. What is the functional group in amine? In amide?

2. What products are formed when amides are hydrolyzed?

3. Draw the condensed structural formulas for each of the following amines and amides:

   a. 1-Propanamine

   b. 2-Methyl-3-hexamine

   c. *N*-Methylpentamide

   d. *N,N*-Dimethylbenzamide

Date ________________ Name ________________________________

Section ________________ Team ________________________________

Instructor ________________ ________________________________

Report Sheet for Lab 22: Amines and Amides

A. Structure, Classification, and Solubility of Amines

	Aniline	*N*-Methylaniline	Triethylamine
Condensed structural formula			
Classification (1°, 2°, 3°)			
Odor			
Solubility in water			
pH			

Questions and Problems

1. In the discussion, the structures are given for histamine and methamphetamine. Give the amine classification of each of these compounds.

2. What type of compound accounts for the "fishy" odor of fish?

3. Explain why amines are basic.

B. Neutralization of Amines with Acid

	Aniline	*N*-Methylaniline	Triethylamine
Solubility after adding HCl			
Change in odor after adding HCl			

Equations for the neutralization of amines with HCl

Aniline + HCl	
N-methylaniline + HCl	
Triethylamine + HCl	

Questions and Problems

4. How does lemon juice remove the odor of fish?

5. Write an equation for the reaction of butylamine with HCl.

C. Amides

1. Amidation

a. Acetic acid + ammonia	
b. Aniline + ammonia	

	Acetamide	Benzamide
2. Odor		
3. Solubility		

D. Hydrolysis of an Amide

1. Acid Hydrolysis

a. Acetamide	
b. Benzamide	

	Acetamide	Benzamide
2. Odor		

	Acetamide	Benzamide
3. Change in color of red litmus paper		
4. Odor after adding NaOH		

5. Base Hydrolysis

a. Acetamide	
b. Benzamide	

Questions and Problems

6. You have unknowns that are a carboxylic acid, an ester, and an amine. Describe how you would distinguish among them.

LAB 23

Amino Acids

A. Amino Acids

Amino acids are used in our bodies to build tissues, enzymes, skin, and hair. The *essential amino acids,* which comprise roughly half of the naturally occurring proteinaceous (protein-containing) amino acids, must be obtained from proteins in the diet since they cannot be synthesized by the body. Amino acids are all similar in structure because each possesses an ionizable amino group ($–NH_3^+$) and an ionizable carboxylate group ($–COO^-$). Individual amino acids have different organic R groups attached to the alpha (α) carbon atom. Variations in these R groups determine whether an amino acid is nonpolar (hydrophobic), polar (hydrophilic), acidic, or basic.

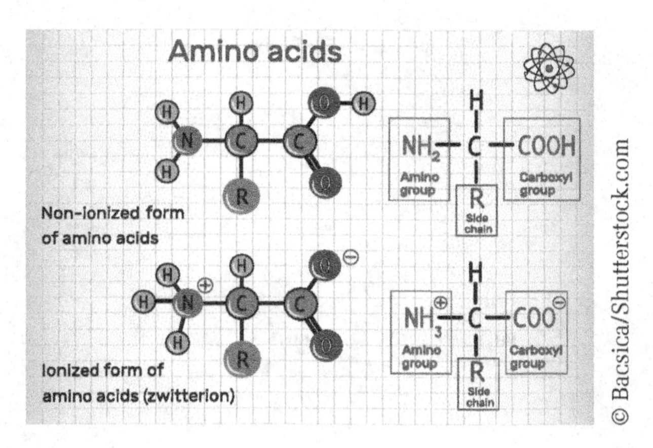

Some R groups contain only carbon and hydrogen atoms, making these amino acids nonpolar and hydrophobic (water-fearing), while other R groups contain –OH or –SH atoms, providing a polar area making them hydrophilic ("water-loving"). Still, other amino acids have R groups that are acidic carboxylic acids or basic amino groups. The R groups of the 20 proteinaceous acids and their chemical characteristics are given below:

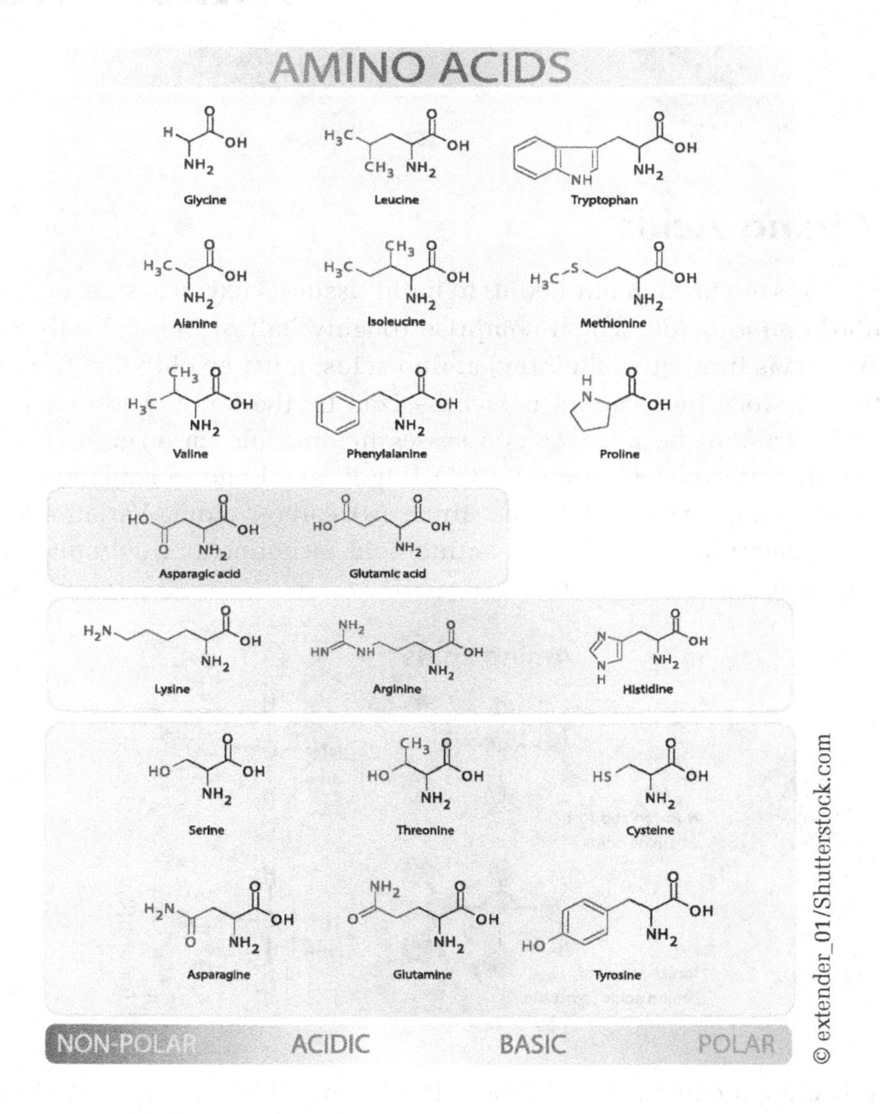

Ionization of Amino Acids

In low pH acidic solutions, the ionized amino group *accepts* a proton (H^+) to form an ion with a positive charge. When placed in a high pH basic solution, the ionized amino acid *donates* a proton (H^+) to form an ion with a negative charge.

Table 23.1 Ionized Forms of Polar and Nonpolar Neutral Amino Acids

Solution	pH < pI (Acidic)	pH = pI	pH > pI (Basic)
Change in H^+	$[H_3O^+]$ increases	No change	$[H_3O^+]$ decreases
Carboxylic Acid/carboxylate	$-COOH$	$-COO^-$	$-COO^-$
Ammonium/amino	$-NH_3^+$	$-NH_3^+$	$-NH_2$
Overall charge	1 +	0	1 −

B. Chromatography of Amino Acids

Chromatography is used to separate and identify the amino acids in a mixture. Small amounts of amino acids and an unknown are placed along one edge of Whatman number 1 paper, which makes the chromatogram. The chromatogram is then placed in a container with solvent. With the paper acting like a wick, the solvent flows up the chromatogram, carrying amino acids along with it. Amino acids that are more soluble in the solvent will move higher up on the paper. Those amino acids that are attracted more to the paper will remain closer to the line of origin. Upon removing and drying the chromatogram, the amino acids can be visualized by spraying the dried chromatogram with ninhydrin. The distance that each amino acid travels from the starting point is measured and the R_f values are calculated using the following formula:

$$R_f = \frac{\text{Distance traveled by an amino acid}}{\text{Distance traveled by the solvent}}$$

To identify an unknown amino acid, its R_f value and color with ninhydrin are compared to the R_f values and colors of the known amino acids. The amino acids present in an unknown mixture of amino acids may be separated and identified in this way.

Experimental Procedures

A. Amino Acids

Materials: Organic model kits or prepared models.

1. Use an organic model kit or observe models of glycine, L-alanine, and L-serine. Draw their condensed structural formulas in the ionized form.
2. Indicate whether each of the amino acids is polar or nonpolar.

3. Draw the condensed structural formulas of glycine, L-ala nine, and L-serine in an acidic solution.
4. Draw the condensed structural formulas of glycine, L-ala nine, and L-serine in a basic solution.

B. Chromatography of Amino Acids

Materials: 600-mL beaker, plastic wrap, nitrile gloves, Whatman chromatography paper number 1 (12 cm × 24 cm), 7 toothpicks or capillary tubes, drying oven (80°C) or hair dryer, metric ruler, stapler, 1% amino acid solutions of L-phenylalanine, L-alanine, glycine, L-serine, L-lysine, L-aspartic acid, and an unknown amino acid, chromatography solvent (isopropyl alcohol, 0.5 M NH_4OH), and 0.2% ninhydrin spray.

Keep fingers off the chromatography paper since amino acids from the skin may be transferred; use pencil, as ink from a pen will dissolve in the solvent and ruin the results.

Using forceps, nitrile gloves, or a paper towel, pick up a piece of Whatman number 1 chromatography paper.

Draw a line using pencil about 2 cm from the long edge of the paper; this will be the line of origin. Mark off 7 equally spaced points about 3 cm apart along the line and label each with the 3-letter abbreviation for each of the amino acids (see Figure 23.1).

Use either the toothpicks or capillary tubes as applicators to make a small spot (about the size of the letter **o**) with each of the 1% amino acid solutions, lightly touching the tip to the paper. *Always return the applicator to the same amino acid solution.* Dry the spots with either a hair dryer or allowing them to air dry. After the spots dry, repeat the application process for the amino acids twice more, for a total of three applications.

Figure 23.1 Chromatography paper labeling.

Prepare the solvent by mixing 10 mL of 0.5 M NH_4OH and 20 mL of isopropyl alcohol.

Pour the solvent into the 600-mL beaker to a depth of about 1 cm, but not over 1.5 cm. The height of the solvent must not exceed the height of the line of origin on the chromatography paper.

Cover the beaker tightly with plastic wrap; this is your chromatography tank.

Roll the chromatography paper with the amino acids into a cylinder and staple the edges *without overlapping. The edges should not touch.*

Slowly lower the cylinder into the solvent of your chromatography tank with the row of the amino acids at the bottom, making sure that the paper does not touch the sides of the beaker (see Figure 23.2).

Figure 23.2 Chromatography tank.

Cover the beaker with the plastic wrap and leave the tank and paper undisturbed.

Let the solvent flow up the paper until the solvent front is 2 to 3 cm from the top edge of the paper; this may take 45 to 60 minutes. *Do not let the solvent run over the top of the paper.*

Carefully remove the paper from the tank. Take the staples out carefully and spread the chromatogram out on a paper towel. *Immediately* mark the solvent line using a ruler and pencil.

Allow the chromatogram to dry completely. A hair dryer or drying oven at about 80°C may be used to speed up the drying process.

Dispose of the solvent in the organic waste as directed by your instructor.

Spray the paper lightly but evenly with ninhydrin. Distinct-colored spots should appear as the ninhydrin reacts with the amino acids in it.

Caution: Do not breath fumes or get spray on your skin.

Calculations

1. Staple the original chromatogram to one member of your lab group's report sheet.
2. Outline each spot with a pencil. Place a dot at the center of each spot. Record the color of each spot.
3. Measure the distance in centimeters from the line of origin to the solvent front to obtain the distance traveled by the solvent and record.
4. Measure the distance in centimeters from the line of origin to the center dot of each spot and record.
5. Calculate and record the R_f values for the known amino acid samples and the unknown.
6. Compare the color and R_f values produced by the unknown amino acid to those of the known samples. Identical amino acids will give similar R_f values and form the same color with ninhydrin. Identify the unknown amino acid.

Date ________________ Name ______________________________

Section ________________ Team ______________________________

Instructor ________________ ______________________________

Prelab Study Questions

1. What two functional groups are in all amino acid?

2. How does an R group determine if an amino acid is acidic, basic, or nonpolar?

3. In a chromatography experiment, a student calculated an R_f value of 0.70 for L-alanine and 0.91 for L-leucine. Which amino acid traveled higher on the chromatography paper? Explain your reasoning.

4. In a chromatography experiment, a student calculated that the solvent front was 6.8 cm above the line of origin. L-arginine traveled a distance of 4.9 cm and glycine traveled 3.4 cm. What R_f values would the student calculate for L-arginine and glycine?

Date ________________ Name ________________________________

Section ________________ Team ________________________________

Instructor ________________ ________________________________

Report Sheet for Lab 23: Amino Acids

A. Amino Acids

1. Condensed structural formulas (ionized) of amino acids		
Glycine	L-Alanine	L-Serine
2. Polar or nonpolar?		
3. Condensed structural formulas of amino acids in an acidic solution		
4. Condensed structural formulas of amino acids in a basic solution		

Questions and Problems

1. Write the structure of the ionized form of L-phenylalanine.

B. Chromatography of Amino Acids

1. Place your chromatogram here:

Calculation of R_f Values

Amino Acid	2. Color	3. Distance (cm) Solvent Traveled	4. Distance (cm) Amino Acid Traveled	5. R_f Value
L-Phenylalanine				
L-Alanine				
Glycine				
L-Serine				
L-Lysine				
L-Aspartic acid				
Unknown				

LAB 24

Peptides and Proteins

A. Peptides

Dipeptides form when two amino acids bond together. An amide, or peptide, bond forms between the carboxylic acid of one amino acid and the amino group of the next amino acid, with the loss of H_2O (a dehydration or condensation reaction). When naming peptides, each amino acid beginning from the N-terminus (nitrogen-containing amino group) has the *-ine* (or *-ic acid*) ending in its name replaced with *-yl*, while the last amino acid in line at the C-terminus (carbon-containing carboxylic acid group) retains its full name. For example, a dipeptide consisting of Glycine (Gly, G) at the N-terminus and Alanine (Ala, A) at the C-terminus would be called Glycylalanine (Gly-Ala, GA).

Formation of a peptide bond

$$^{+}H_3N-\underset{R}{\overset{H}{C}}-\overset{O}{\overset{\|}{C}}-O^- + {}^{+}H_3N-\underset{R}{\overset{H}{C}}-\overset{O}{\overset{\|}{C}}-O^- \longrightarrow {}^{+}H_3N-\underset{R}{\overset{H}{C}}-\overset{O}{\overset{\|}{C}}-\underset{H}{N}-\underset{R}{\overset{H}{C}}-\overset{O}{\overset{\|}{C}}-O^- + H_2O$$

Peptide bond

Hydrolysis is the reverse reaction in which water adds to the peptide bond to yield individual amino acids. When the tripeptide Alanylglycylserine (Ala-Gly-Ser, AGS) hydrolyzes, for example, it gives the amino acids Alanine (Ala, A), Glycine (Gly, G), and Serine (Ser, S).

Proteins

Several amino acids joined together by peptide bonds form a polypeptide. Fifty or more amino acids in a peptide chain with biological activity are considered a protein. Proteins make up many vital tissues in the body, including skin, muscle, cartilage, hair, fingernails, enzymes, and hormones. Proteins have specific structures, determined by their amino acid sequence. Peptide bonds joining amino acids together comprise the *primary* protein structure, while *secondary* protein structures include the alpha (α) helix formed by the coiling of peptide chains and β-pleated sheets formed between protein strands. Many hydrogen bonds between the oxygen atoms of the carbonyl groups and the hydrogen atoms of the amide groups hold the secondary structures in place. In *tertiary* protein structure, interactions between side groups like ionic bonds, salt bridges, disulfide bonds, and hydrophilic interactions give it a compact shape; indeed, these tertiary structures are evident in globular proteins' spherical shape. Finally, similar interactions between two or more tertiary units produce the *quaternary* structure of many active proteins.

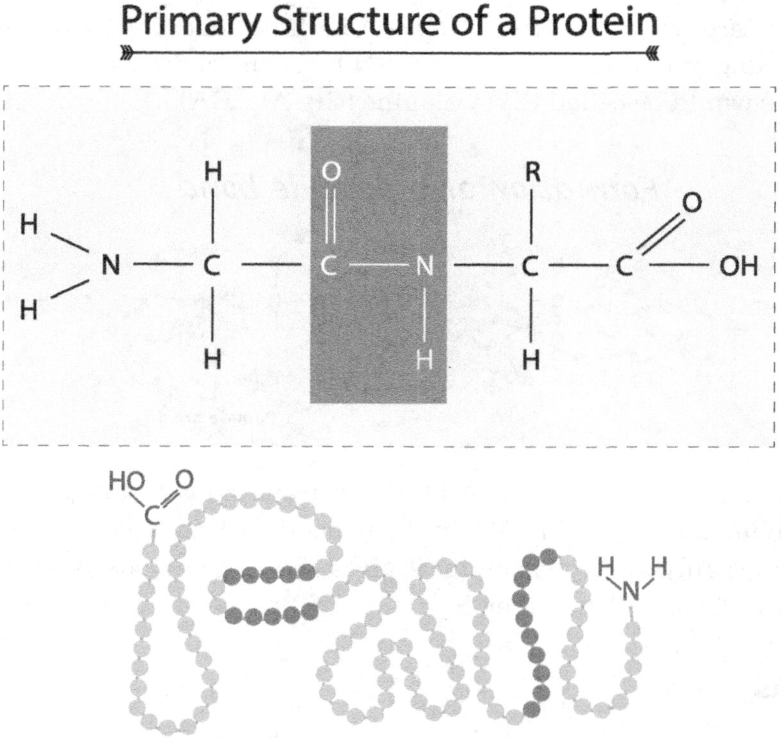

Primary Structure of a Protein

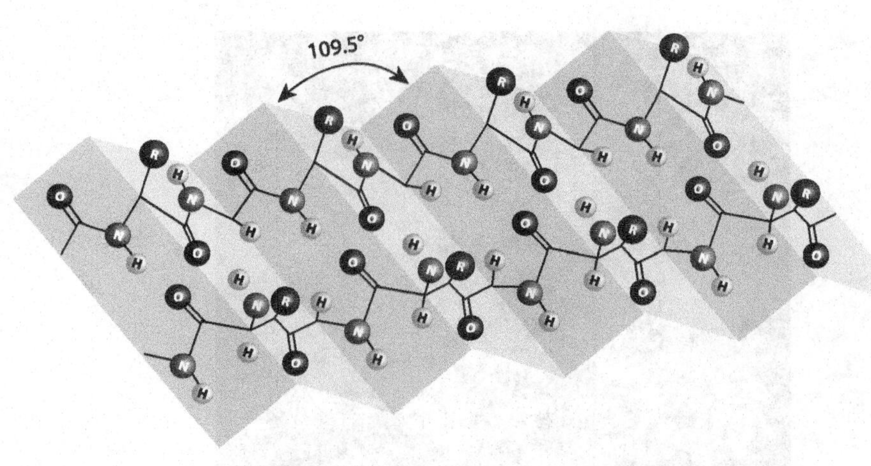

Secondary structure of a β-pleated sheet

Tertiary structure of myoglobin

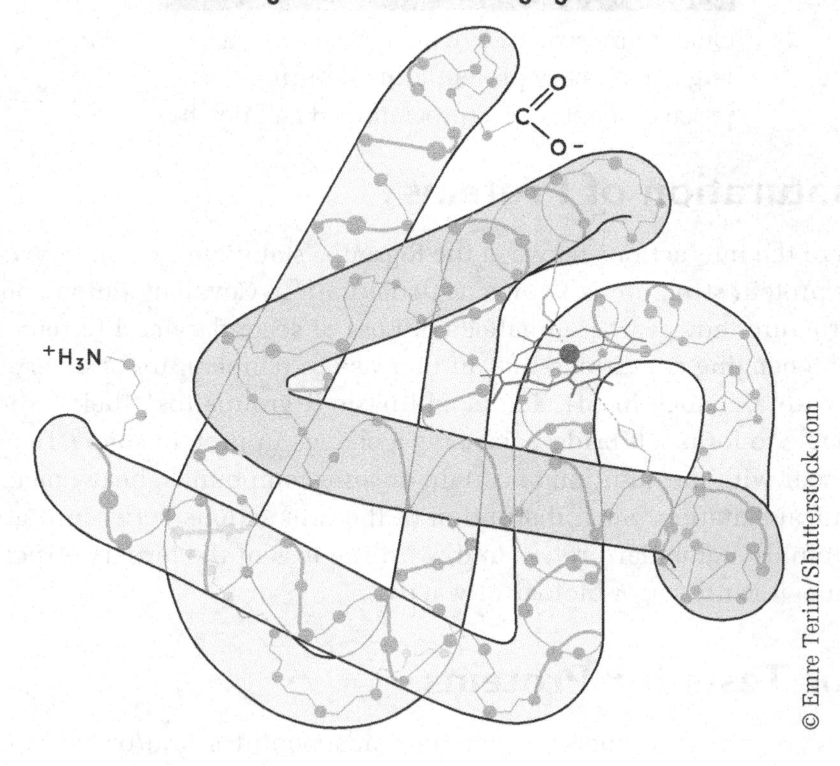

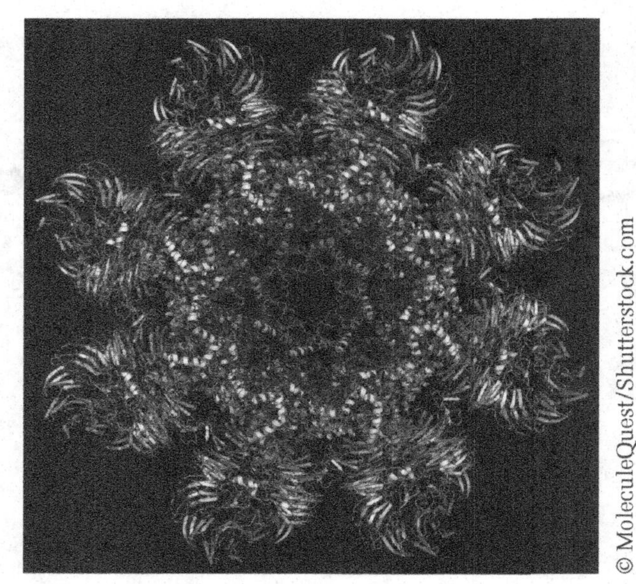

Quaternary structure of an apoptosome, a large quaternary protein formed during the process of apoptosis (programmed cell death)

B. Denaturation of Proteins

Disruption of the interactions between the R groups stabilizing secondary, tertiary, or quaternary protein structure is known as denaturation. Covalent amide bonds of the primary structure, however, are unaffected. Loss of secondary and tertiary structure occurs with changing conditions like an increase in temperature or a very acidic or basic pH. With a change in pH, acidic and basic R groups lose their ionic charges and are unable to form salt bridges, causing a change in protein shape. Denaturation may also occur with the addition of certain organic compounds, heavy metal ions, or via mechanical agitation. Such disruption of the interactions between R groups, for instance, unfolds a globular protein and, with the loss of the tertiary structure of its overall shape, it is no longer biologically active.

C. Color Tests for Proteins

Certain tests give color products with amino acids, peptides, and/or proteins. The results of the test can be used to detect certain groups or types of bonds within proteins or amino acids.

Biuret test: The biuret test is positive for a peptide or protein with three (tripeptide) or more peptide bonds. With the *biuret test*, the blue color of a basic solution of Cu^{2+} changes to a violet color when a tripeptide or larger peptide is present; individual amino acids and dipeptides do not react with the reagent and the solution remains blue (negative).

Ninhydrin test: The ninhydrin test is employed to detect amino acids and most proteins. With this test, most amino acids yield a blue-violet color, while proline and hydroxyproline produce a yellow color.

Xanthoproteic test: This test is specific for amino acids containing an aromatic ring. Concentrated nitric acid reacts with the side chains of phenylalanine, tyrosine, and tryptophan to give nitro-substituted benzene rings that appear as yellow-colored products.

Experimental Procedures

A. Peptides

Materials: Organic model set.

1. Use the model set or observe models of the ionized forms of the amino acids glycine and serine, and draw their ionized, condensed structural formulas.
2. By removing the components of water (H–OH) from the amino and carboxylate groups, draw the condensed structural formulas of the dipeptides glycylserine and serylglycine.
3. Draw the condensed structural formulas of the reactants and products for the hydrolysis of the dipeptide serylglycine by breaking the peptide bond and adding the components of H_2O.
4. Draw the condensed structural formula for the tripeptide serylglycylalanine.
5. Write the 1- and 3-letter abbreviations for this tripeptide.

B. Denaturation of Proteins

Materials: 5 test tubes, test tube holder, 10-mL graduated cylinder, 1% albumin solution, 10% HNO_3, 10% NaOH, 95% ethyl alcohol, 1% $AgNO_3$, propane torch.

Place 2 to 3 mL of albumin solution in each of the 5 test tubes and label 1 to 5.

For each test, record your observations and give a reason for the changes in the albumin protein.

1. **Heat**: Using a test tube holder, heat the albumin solution over a low flame.
2. **Acid**: Add 2 mL of 10% HNO_3.
3. **Base**: Add 2 mL of 10% NaOH.
4. **Alcohol**: Add 4 mL of 95% ethyl alcohol solution.
5. **Heavy metal ions**: Add 10 drops of 1% $AgNO_3$. Caution, $AgNO_3$ stains the skin.

C. Color Tests for Proteins

Materials: 12 test tubes, test tube rack, 10-mL graduated cylinder, stirring rod, 2 beakers able to accommodate (1) a boiling water bath and (2) a cold water bath, pH paper, spatula or spoonula, 1% glycine solution, tyrosine powder, 1% albumin solution, gelatin powder, 0.2% ninhydrin solution, 10% HNO_3, 10% NaOH, 0.1 M $CuSO_4$ (biuret), red litmus paper.

1. **Biuret test**: Place 2 mL of glycine solution in a test tube.
   Add 2 mL of 10% NaOH and stir.
   Add 5 drops of biuret reagent ($CuSO_4$) and stir.
   Record the color.
   Positive biuret test: The formation of a pink-violet color indicates the presence of a protein with three or more peptide bonds; if such a protein is not present, the blue color of the copper(II) sulfate will remain (negative result).

   Repeat the above procedure for the biuret test using a small amount (the tip of a spatula or spoonula) of tyrosine, gelatin, (each dissolved in 2 mL of distilled water), and 2 mL of albumin.
2. **Ninhydrin test**: Place 2 mL of glycine solution in a test tube.
   Add 1 mL of 0.2% ninhydrin solution.
   Place the test tube in a boiling water bath for 4 to 5 minutes.
   Record your observations.

Positive ninhydrin test: Look for the formation of a blue-violet color.

Repeat the above procedure for the ninhydrin test using a small amount (the tip of a spatula or spoonula) of tyrosine, gelatin (each dissolved in 2 mL of distilled water), and 2 mL of albumin.

Record the colors obtained for these remaining samples.

3. **Xanthoproteic test**: Place 2 mL of glycine solution in a test tube.

   Add 10 drops of 10% HNO_3 to the sample.

   Place the test tube in a boiling water bath and heat for 3 to 4 minutes.

   Remove the test tube, place it in a cold water bath and allow to cool for 3 to 4 minutes.

   Carefully add 10% NaOH dropwise until the solution is just basic (turns red litmus paper blue). This may require 2 to 3 mL of NaOH. ***Caution: heat will be evolved.***

   Positive xanthoproteic test: Look for the formation of a yellow-orange color, which may vary in intensity.

   Repeat the above procedure for the xanthoproteic test using a small amount (the tip of a spatula or spoonula) of tyrosine, gelatin (each dissolved in 2 mL of distilled water), and 2 mL of albumin.

   Record the colors obtained for these remaining samples.

Date ______________ Name ______________________________

Section ______________ Team ______________________________

Instructor ______________ ______________________________

Prelab Study Questions

1. What is a peptide bond?

2. How does the primary structure of a protein differ from the secondary structure?

3. Write the 1- and 3-letter abbreviations for the 6 tripeptides possible from Ala, Pro, and Ser.

4. Draw the condensed structural formula of the dipeptide Phe-Val (FV).

Date ________________ Name ________________________________

Section ________________ Team ________________________________

Instructor ________________ ________________________________

Report Sheet for Lab 24: Peptides and Proteins

A. Peptides

1. Condensed structural formulas of glycine and serine

Glycine	Serine

2. Condensed structural formulas of dipeptides

Glycylserine	Serylglycine

3. Condensed structural formulas of the reactants and products for the hydrolysis of serylglycine

4. Condensed structural formula for the tripeptide alanylglycylserine

5. Abbreviations for the above tripeptide

3-letter abbreviation	1-letter abbreviation

B. Denaturation of Proteins

Treatment	Observations	Reason for Changes
Heat		
Acid		
Base		
Alcohol		
Heavy metal ions		

Questions and Problems

1. Why are heat and alcohol used to disinfect medical equipment?

2. Why is milk given to someone who accidentally ingests a heavy metal ion such as silver or mercury?

C. Color Tests for Proteins

Observations of Color Tests			
Sample	Biuret	Ninhydrin	Xanthoproteic
Glycine			
Tyrosine			
Gelatin			
Albumin			

Questions and Problems

3. After working with HNO_3, a student noticed that she had a yellow spot on her hand. What might be the reason?

4. Which samples gave a negative biuret test? Why?

5. What functional group gives a positive xanthoproteic test?

6. What tests could you use to determine whether an unlabeled test tube sample contained an amino acid or a protein?

LAB 25

Enzymes

Enzymes catalyze reactions in biological systems, speeding them up while operating at relatively mild conditions of temperature and pH. As catalysts, enzymes lower the activation energy (E_a or threshold energy) for chemical reactions (see Figure 25.1). With less energy required to convert reactant molecules to products, the rate of catalyzed biochemical reactions is greatly increased compared to the rate of the uncatalyzed reaction. In fact, the rates of enzyme-catalyzed reactions are much, much faster than the rates of the uncatalyzed reactions. Some enzymes can increase the rate of a biological reaction by a factor of a billion, a trillion, or even one hundred million trillion compared to the rate of the uncatalyzed reaction. For example, the enzyme in blood called carbonic anhydrase converts carbon dioxide and water to carbonic acid. This enzyme catalyzes the reaction of about 35 million molecules of CO_2 every minute.

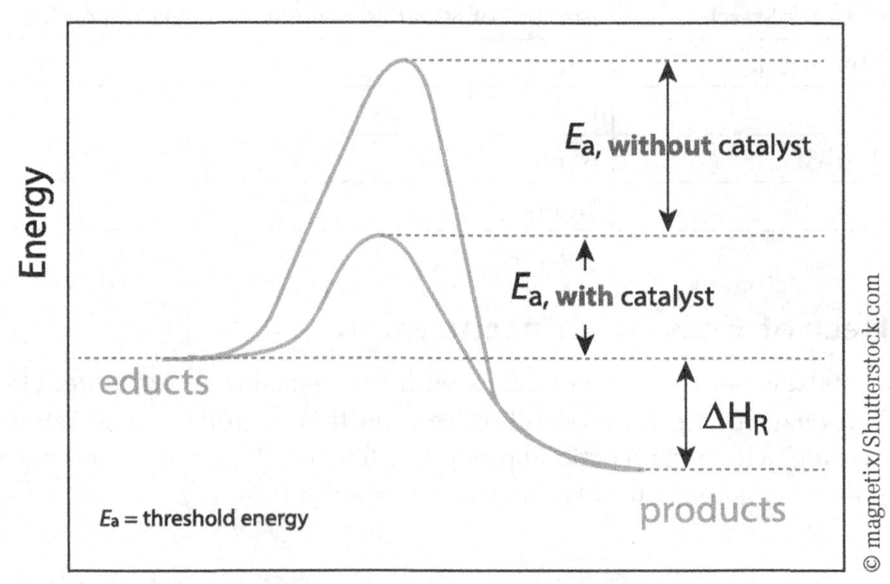

Figure 25.1 Activation (or threshold) energy diagram.

Testing Enzyme Activity

In this experiment, the enzyme amylase, which catalyzes the hydrolysis of amylose, will be used. In the presence of amylase, a starch sample will undergo hydrolysis to give smaller polysaccharides, dextrins, maltose, and eventually glucose.

$$\text{Starch (amylose)} \xrightarrow{\textit{Amylase}} \text{Smaller polysaccharides, dextrins, maltose} \xrightarrow{\textit{Amylase}} \text{Glucose}$$

Visual Color Reference

As you proceed with each experiment, you will check enzyme activity by adding iodine to the starch mixture. When enzyme activity is high, the time required for the starch to be hydrolyzed will be very short; when enzyme activity is slowed down or inactive, the blue-black color of the iodine will persist for a longer time. By observing the rate of disappearance of starch, you can assess the relative amount of enzyme activity using Table 25.1:

Table 25.1 Iodine Test Color and Enzyme Activity Comparison Table

Iodine Test for Starch	Amount of Starch Remaining	Enzyme Activity
Dark blue-black	All	0
Blue	Most	1
Light brown	Some	2
Gold	None	3

A. Effect of Enzyme Concentration

During catalysis, an enzyme combines with the reactant, or *substrate*, of a reaction to give an *enzyme–substrate* complex. To form this complex, the substrate fits into the *active site*, where the reaction takes place. The *product(s)* is/are released, and the original enzyme is ready to catalyze another reaction.

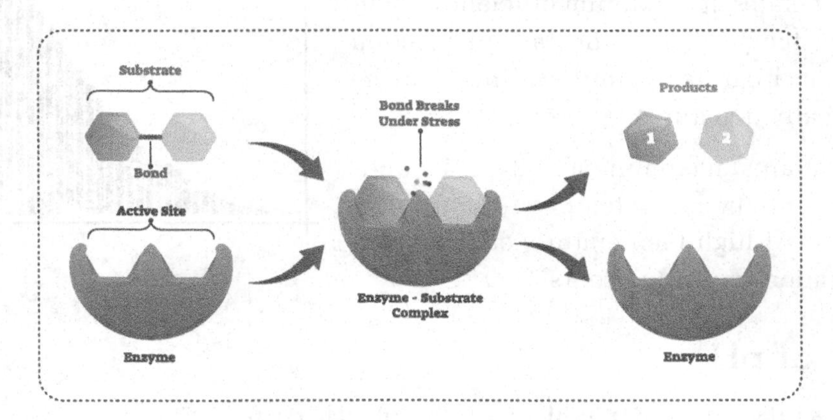

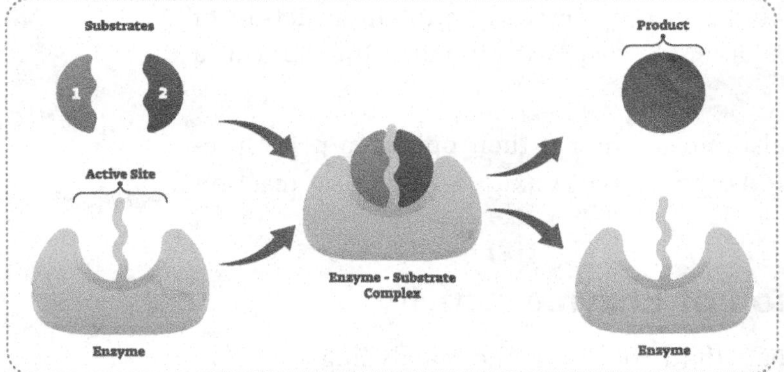

If the enzyme concentration is increased while the substrate concentration remains constant, the rate of the reaction increases. With more enzyme present, more substrate molecules can react.

Increasing the enzyme concentration increases the rate of the reaction.

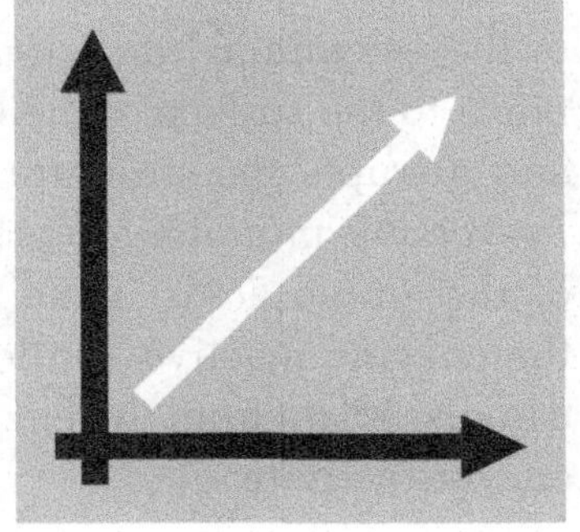

B. Effect of Temperature

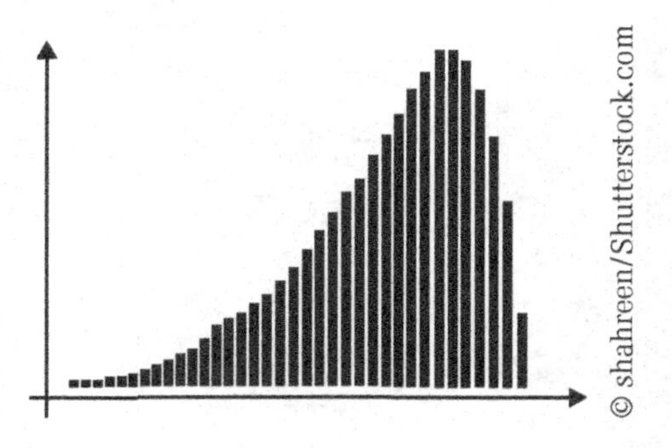

Optimum temperature is the temperature at which an enzyme operates at maximum efficiency. Typically, at low temperatures, the rate of reaction slows and at high temperatures, the enzyme protein typically denatures.

An enzyme attains maximum activity at its optimum temperature (*x* axis = temperature, *y* axis = reaction rate). At high temperature, enzyme protein denaturation typically occurs.

C. Effect of pH

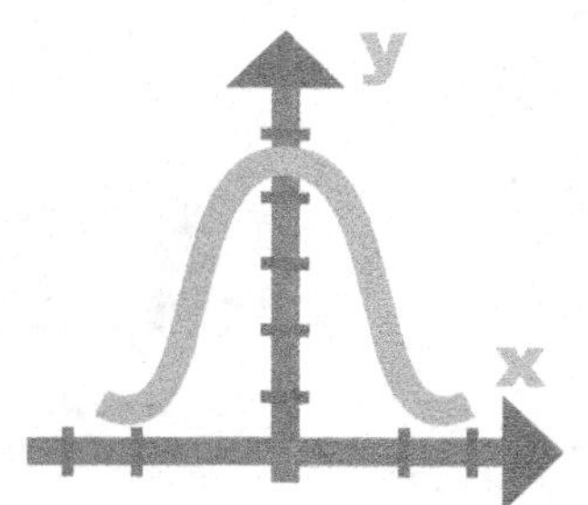

An enzyme is also most active at its optimum pH. At pH values below and above optimal, the protein structure of the enzyme is altered, severely reducing the enzyme's activity.

Enzymes are also most active at their optimum pH, represented by the top of the curve (*x* axis = pH, *y* axis = reaction rate).

D. Inhibition of Enzyme Activity

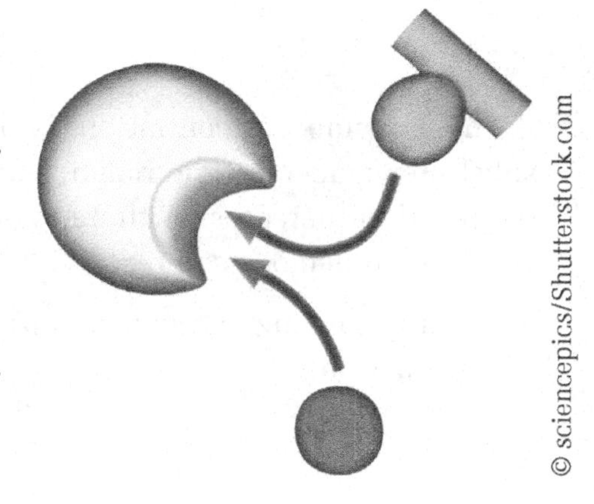

Substances that limit or stop the catalyzing activity of an enzyme are known as *inhibitors*. A *competitive inhibitor* competes for the active site of the enzyme, while a *noncompetitive inhibitor* binds to the surface of the enzyme at another site and disrupts the structure of the active site. An *irreversible inhibitor* forms bonds with side chains of the amino acids in the active site, rendering the enzyme inactive, while a *reversible inhibitor* can dissociate from the enzyme, thereby restoring enzyme activity.

A competitive inhibitor has a similar structure to that of the substrate, fitting into the active site and competing with the substrate.

Experimental Procedures

A. Effect of Enzyme Concentration

Materials: 4 test tubes, test tube rack, thermometer, stirring rod, 250-mL beaker, 4 droppers, spot plate, hot plate, amylase preparation, 1% starch solution, iodine reagent, 5- or 10-mL graduated cylinder.

Preparation of warm water bath: Fill the beaker roughly 1/2 to 2/3 full of tap water, warm it on the hot plate to about 37°C, and try to maintain that temperature.

Testing for Enzyme Activity

Add 5 mL of 1% starch to 4 separate test tubes and label them 1 to 4. Test tube 1 without enzyme will be the reference. Place all test tubes in the warm water bath and after 5 minutes add the following number of drops of amylase to the other test tubes and mix.

	Test Tube 1	Test Tube 2	Test Tube 3	Test Tube 4
Starch	5 mL	5 mL	5 mL	5 mL
Amylase	None	3 drops	6 drops	9 drops

1. Using 4 separate droppers, immediately (0 min) transfer 5 drops of each reaction mixture in test tubes 1 to 4 to the spot plate and add 1 drop of iodine reagent to each solution, recording your observations.

   Use the visual color reference (Table 25.1) to assess enzyme activity.

   Rinse the spot plate.
2. After 5 minutes, test the samples for starch again and record your results.
3. After 10 minutes, test the samples for starch again and record your results.

B. Effect of Temperature

Materials: 6 test tubes, 3 rubber test tube stoppers, test tube rack, test tube holder, 4 droppers, 5- or 10-mL graduated cylinder, 3 × 250-mL beakers for water baths as follows: one roughly half-full of ice water (0°C–5°C), one roughly half-full with warm water (37°C), and one roughly half-full with water heated to boiling (100°C), amylase preparation, spot plate, 1% starch solution, iodine reagent.

Preparing the Starch Solution

Pour 5 mL of 1% starch solution into each of the 3 test tubes. Place one test tube in the boiling water bath, one in the warm water bath, and one in the ice-water bath. Allow the test tubes to remain in their respective water baths for 5 minutes so that they reach the bath temperature. Prepare the amylase solutions during this time.

Preparing the Amylase Solution

Pour 1 mL of amylase solution into each of the 3 remaining test tubes. Place one test tube in the boiling water bath, one in the warm water bath, and one in the ice-water bath. Allow the test tubes to remain in their respective water baths for 5 minutes so that they reach the bath temperature.

Mix the Starch and Amylase Solutions

Remove the 2 test tubes from the boiling water bath using the test tube holder and pour the 5 mL of starch into the amylase-containing test tube. Stopper and mix by shaking. Remove stopper and return this test tube to the boiling water bath for 5 minutes.

Remove the 2 test tubes from the warm water bath using the test tube holder and pour the 5 mL of starch into the amylase-containing test tube. Stopper and mix by shaking. Remove stopper and return this test tube to the warm water bath for five minutes.

Remove the 2 test tubes from the ice-water bath using the test tube holder and pour the 5 mL of starch into the amylase-containing test tube. Stopper and mix by shaking. Remove stopper and return this test tube to the ice-water bath for five minutes.

	Test Tube 1	Test Tube 2	Test Tube 3
1% starch	5 mL	5 mL	5 mL
Temperature	0°C	37°C	100°C
Amylase	1 mL	1 mL	1 mL

Testing for Enzyme Activity

1. Transfer 4 drops of the mixture from the boiling water bath to the spot plate. Add 1 drop of iodine reagent. Record the color assess the enzyme activity.
2. Transfer 4 drops of the mixture from the warm water bath to the spot plate. Add 1 drop of iodine reagent. Record the color assess the enzyme activity.
3. Transfer 4 drops of the mixture from the ice-water bath to the spot plate. Add 1 drop of iodine reagent. Record the color assess the enzyme activity.

C. Effect of pH

Materials: 8 test tubes, 4 rubber test tube stoppers, test tube rack, 5 droppers, amylase preparation, 37°C water bath, buffers of pH 2, 4, 7, and 10, spot plate, 1% starch solution, iodine reagent.

Pour 5 mL of 1% starch solution into each of 4 test tubes and place in the 37° C water bath for 5 minutes.

Pour 4 mL, respectively, of pH 2, 4, 7, and 10 into the other 4 test tubes. Add 1 mL of amylase solution to each and place them in the 37°C water bath for 5 minutes.

After 5 minutes, pour each of the 1% starch solutions into each of the pH buffer-amylase test tubes, stopper, and mix by shaking. After mixing, remove stoppers, and return the mixtures to the 37°C water bath for 5 minutes.

	Test Tube 1	Test Tube 2	Test Tube 3	Test Tube 4
1% starch	5 mL	5 mL	5 mL	5 mL
Buffer solution	4 mL of pH 2	4 mL of pH 4	4 mL of pH 7	4 mL of pH 10
Amylase	1 mL	1 mL	1 mL	1 mL

Testing for Enzyme Activity

Remove 5 drops from each of the pH 2, 4, 7, and 10 test tubes, respectively, and place on the spot plate. Add 1 drop of iodine reagent to each, record your observations, and assess enzyme activity.

D. Inhibition of Enzyme Activity

Materials: 6 test tubes, test tube rack, 4 droppers, 3 rubber test tube stoppers, 5- or 10-mL graduated cylinder, amylase preparation, 37°C water bath, spot plate, the following solutions: distilled water, 0.1 M $AgNO_3$, 95% ethanol, 1% starch solution, iodine reagent.

Place 1 mL of amylase preparation into 3 test tubes and label them 1, 2, and 3.

Place 5 mL of 1% starch solution into the other 3 test tubes.

To test tube 1, add 10 drops of distilled water as your reference.

To test tube 2, add 10 drops of 0.1 M $AgNO_3$.

To test tube 3, add 10 drops of 95% ethanol.

Place all 6 test tubes in the 37°C water bath for 5 minutes.

Mix Starch, Inhibitors, and Amylase

After 5 minutes, pour one each of the 1% starch solutions into one each, respectively, of the inhibitor-amylase test tubes, stopper, and mix by shaking. After shaking, remove stoppers, and return the mixtures to the 37°C water bath for 5 minutes.

	Test Tube 1	Test Tube 2	Test Tube 3
1% starch	5 mL	5 mL	5 mL
Inhibitor	10 drops water	10 drops $AgNO_3$	10 drops ethanol
Amylase	1 mL	1 mL	1 mL

Testing for Enzyme Activity

Remove 5 drops from each of the test tubes and place on the spot plate. Add 1 drop of iodine reagent to each, record your observations, and assess enzyme activity.

Date ________________ Name ______________________________

Section ________________ Team ______________________________

Instructor ________________ ______________________________

Prelab Study Questions

1. What is the substrate for the amylase enzyme?

2. What are the products of the amylase action?

3. What happens to most enzymes at high temperature?

4. What happens to an enzyme when a competitive inhibitor binds to it?

Date ________________ Name ________________________________

Section ________________ Team ________________________________

Instructor ________________ ________________________________

Report Sheet for Lab 25: Enzymes

A. Effect of Enzyme Concentration

Time		Test Tube 1	Test Tube 2	Test Tube 3	Test Tube 4
		No amylase	3 drops amylase	6 drops amylase	9 drops amylase
1. 0 minutes	Color				
	Activity				
2. 5 minutes	Color				
	Activity				
3. 10 minutes	Color				
	Activity				

Questions and Problems

1. **a.** At which enzyme concentration was starch hydrolyzed the fastest? Explain.

   b. At which enzyme concentration was starch hydrolyzed the slowest? Explain.

2. Describe the effect of enzyme concentration on enzyme activity.

B. Effect of Temperature

0°C		37°C		100°C	
Color	Activity	Color	Activity	Color	Activity

Questions and Problems

3. What was the optimal temperature for amylose?

4. a. Why would the enzyme activity be different in the 0°C sample versus the 37°C sample?

   b. Why would the enzyme activity be different in the 37°C sample versus the 100°C sample?

C. Effect of pH

pH 2		pH 4		pH 7		pH 10	
Color	Activity	Color	Activity	Color	Activity	Color	Activity

Questions and Problems

5. a. How is amylase activity affected by a low pH? Explain.

   b. How is amylase activity affected by a high pH? Explain.

6. What was the optimum pH for amylase?

7. a. During digestion, the pH of the stomach is 2. What does this indicate about the optimum pH of pepsin, the enzyme that hydrolyzes proteins in the stomach?

   b. What happens to the activity of pepsin when it enters the small intestine, where the pH is 8?

D. Inhibition of Enzyme Activity

Water		$AgNO_3$		Ethanol	
Color	Activity	Color	Activity	Color	Activity

Questions and Problems

8. In which reaction mixture(s) did starch hydrolysis occur?

9. Which substances added to the mixture(s) were inhibitor? Which were not?

10. What are some differences and/or similarities in the type of inhibition caused by heat, acid, base, heavy metal ions, and ethanol on enzyme activity?

LAB 26

Saponification and Soaps

A. Saponification: Preparation of Soap

For centuries, soaps have been made from heating animal fats and lye (NaOH), the lye being obtained from pouring water through wood ashes. This hydrolysis of a fat or oil by an alkaline base like NaOH is known as *saponification*, and the salts of the fatty acids obtained are called *soaps*. The other product of this hydrolysis, glycerol, is soluble in water. Using NaOH produces a solid soap, which can be molded into a desired shape. When KOH is used, a softer liquid soap is produced. Polyunsaturated oils also produce softer soaps; names like "coconut" or "avocado shampoo" reveal the sources of the oils used in the reaction.

Fat or oil + strong base $\rightarrow$ Glycerol + salts of fatty acids (soap)

Glyceryl tristearate + 3NaOH $\rightarrow$ glycerol + 3 Sodium palmitate

The fats most commonly used to make soaps are lard and tallow from animal fat, and coconut, palm, and olive oils from vegetables. Castile soap is made from olive oil; soaps that float contain air pockets, and soft soaps made with KOH instead of NaOH give the potassium salts.

B. Properties of Soaps and Detergents

Soap molecules have a dual nature; the nonpolar carbon chains are hydrophobic and are attracted to nonpolar substances like grease, while the polar heads of the carboxylate salts are hydrophilic and are attracted to water (Figure 26.1).

Nonpolar head (hydrophobic) Polar head (hydrophilic)

Figure 26.1 The dual polarity of a soap (salt of a fatty acid).

Preparation of Soaps

Soaps, typically prepared from fats such as coconut oil, have essential or fragrance oils added to give a pleasant-smelling product. A soap is the salt of a long-chain fatty acid, having a long hydrocarbon-like chain with a carboxylate negative end, combined with a positive sodium or potassium ion. Soaps possess dual properties since parts of the soap molecule have different properties; the sodium or potassium carboxylate end is ionic and very water soluble ("hydrophilic") but insoluble in oils or grease, while the long hydrocarbon end is insoluble in water ("hydrophobic"), but soluble in nonpolar substances like oil or grease. When soap is used, the hydrocarbon tails of the soap molecules dissolve in the nonpolar fats and oils that accompany dirt. The soap molecules then coat the oil or grease and form clusters called *micelles*, in which the hydrophilic salt ends of the soap molecules extend outside, where they can dissolve in water. As a result, small globules of oil and fat coated with soap molecules are pulled into the water and rinsed away. In hard water, where high concentrations of minerals are present, the carboxylate ends of soap react with Ca^{2+}, Fe^{3+}, or Mg^{2+} ions, forming an insoluble substance that we see as a gray line in the bathtub or sink. This does not occur with detergents (Figure 26.2).

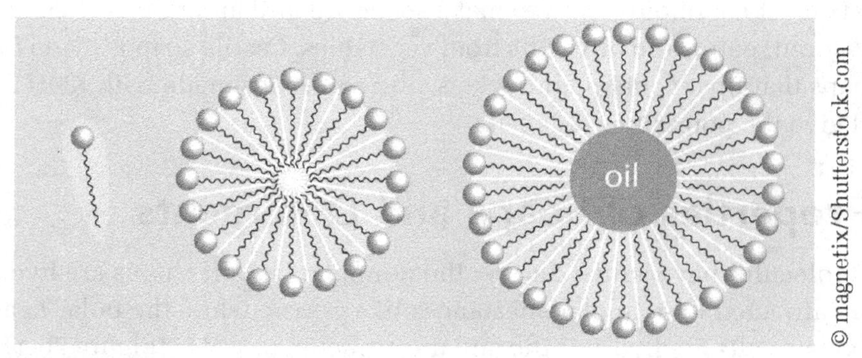

Figure 26.2 The hydrocarbon tails of soap molecules dissolve in grease and oil to form micelles, which are pulled by their ionic heads into the rinse water.

Detergents

Detergents represent synthetic cleaning agents since they are not derived from naturally occurring fats or oils. Their popularity lies in the fact that they do not form insoluble salts with ions, meaning that they work in hard as well as soft water; a typical detergent is sodium lauryl sulfate, also known as sodium dodecyl sulfate.

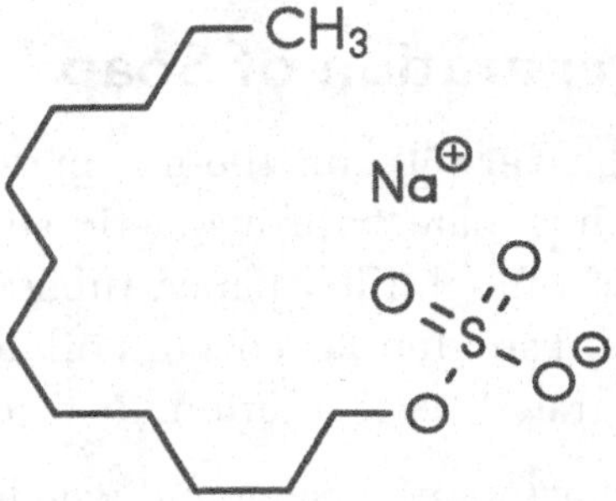

sodium dodecyl sulfate

As detergents replaced soaps for cleaning, it was discovered that they were not degraded in sewage treatment plants and large amounts of foam appeared in streams and lakes that became polluted with detergents. Biodegradable detergents like lauryl glucoside eventually replaced the nonbiodegradable ones.

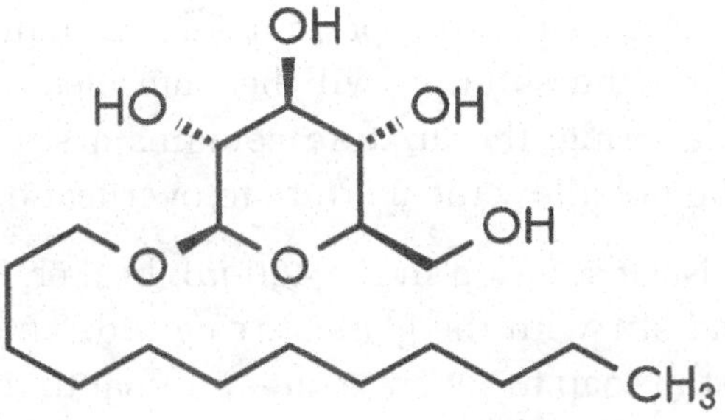

Lauryl glucoside

Commercial detergents, however, also contain phosphate compounds along with brighteners and fragrances; phosphates accelerate the growth of algae in lakes, known as algal blooms, which deplete the dissolved oxygen in the water. As a result, fish and other living organisms die. The use of phosphates in detergents is now decreasing though and nonphosphate replacements have been developed.

Experimental Procedures

A. Saponification: Preparation of Soap

Materials: 150- or 400-mL beaker, 50- or 100-mL graduated cylinder, glass stirring rod, electronic balance, hot plate with magnetic stirring bar, magnetic stirrer, 2 large watch glasses, Büchner funnel, filter paper, rubber disk, nitrile gloves, safety glasses or goggles, fat (lard, solid shortening, coconut oil, olive or other vegetable oil), 95% ethanol, 20% NaOH, saturated NaCl solution, distilled water.

Place the magnetic stirrer in the 150-mL beaker and weigh on the electronic balance. Tare or zero the balance and add about 5 g of fat or oil to the beaker.

Add 15 mL of 95% ethanol and 15 mL of 20% NaOH. ***Use care when pouring NaOH.*** Wear nitrile gloves.

Caution: Oil and ethanol will be hot and may splatter or catch fire. Keep a large watch glass nearby to smother any flames. NaOH is caustic and can cause permanent eye damage. Wear safety glasses or goggles at all times.

Place the beaker on the hot plate, turn on the hot plate's magnetic stirring bar, cover with the second watch glass, and heat to a gentle boil for 30 minutes, adding 5-mL portions of a 1:1 ethanol–water mixture periodically to maintain volume, or until saponification is complete (the solution will become clear with no separation of layers). Be careful of splattering; the mixture contains a strong base. If foaming is excessive, *reduce* heat. Do not allow the mixture to overheat or to char.

Pour 50 mL of saturated NaCl solution in the 400-mL beaker. Pour the soap solution into this salt solution and stir with the glass stirring rod. This process is known as "salting out" and causes the soap to separate out and float to the surface.

Collect the solid soap using a Büchner funnel and filter paper (see Figure 26.3).

Wash the soap with two 10-mL portions of cold distilled water, pulling air through the product to dry it further.

Place the soap curds obtained on one of the watch glasses, using nitrile gloves to handle the soap if necessary. ***Handle with care: The soap may still contain NaOH, which can irritate the skin.***

Describe the soap that you prepared. *Save the soap for **Part B**.*

Figure 26.3 Apparatus for suction filtration with Büchner funnel.

B. Properties of Soaps and Detergents

Materials: 3 test tubes, 3 rubber test tube stoppers, 3 droppers, 3 small beakers, 50- or 100-mL graduated cylinder, glass stirring rod, wax pencil, laboratory-prepared soap from **Part A**, commercial soap product, detergent, pH paper, oil, 0.1 M $CaCl_2$, 0.1 M $MgCl_2$, and 0.1 M $FeCl_3$ solutions.

Using the 3 small beakers, prepare solutions of the soap made in **Part A**, the commercial soap, and the detergent, by dissolving about 1 g of each in 50 mL of distilled water. If the commercial soap is a liquid, use 20 drops.

pH Test

Place 10 mL of each solution in separate test tubes and label.

Dip the glass stirring rod into one of the solutions, then touch the end of the rod to the pH paper, recording the pH. Rinse the stirring rod and repeat for the other two solutions. *Save the test tubes for the **Foam test**.*

Foam Test

Stopper each of the test tubes from the **pH test** and shake for 10 seconds. The soap should form a layer of suds or foam. Record your observations. *Save the test tubes for the **Reaction with oil**.*

Reaction with Oil

Remove the stoppers and add 5 drops of oil to each test tube from the **Foam test**. Stopper each test tube again and shake each one for 10 seconds. Record your observations, comparing the sudsy layer in each test tube to the sudsy layers in the **Foam test**. *Save the test tubes for the **Hard water test**.*

Hard Water Test

Remove the stoppers and add 20 drops of 0.1 M $CaCl_2$ to the first test tube, 20 drops of 0.1 M $MgCl_2$ to the second test tube, and 20 drops of 0.1 M $FeCl_3$ to the third test tube. Stopper each test again and shake each one for 10 seconds. Compare the foamy layer in each of the test tubes to the sudsy layers in the **Foam test** and record your observations for each.

Date ______________ Name ______________________

Section ______________ Team ______________________

Instructor ______________ ______________________

Prelab Study Questions

1. What happens when a fatty acid is reacted with NaOH?

2. Why is ethanol added to the reaction mixture of fat and base in the making of soap?

3. Why is the product of saponification a salt?

4. Draw the condensed structural formulas in the equation for the saponification of tristearin with KOH.

Date ________________ Name ________________________________

Section ________________ Team ________________________________

Instructor ________________ ________________________________

Report Sheet for Lab 26: Saponification and Soaps

A. Saponification: Preparation of Soap

Describe the appearance of the soap that you prepared.

B. Properties of Soaps and Detergents

Tests	Lab Soap	Commercial Soap	Detergent
1. pH			
2. Foam			
3. Oil			
4. 0.1 M $CaCl_2$			
0.1 M $MgCl_2$			
0.1 M $FeCl_3$			

Questions and Problems

1. Which of the solutions in **Part B** were basic?

2. How do soaps made from vegetable oils differ from soaps made from animal fat?

3. How does soap remove an oil spot?

4. Draw the condensed structural formulas in the equation for the saponification of trimyristin with KOH.

Common Ion Table

Common Multivalenced Cations		
Element	Name	Symbol
Chromium	Chromium (II) or chromous	Cr^{2+}
Chromium	Chromium (III) or chromic	Cr^{3+}
Cobalt	Cobalt (II) or cobaltous	Co^{2+}
Cobalt	Cobalt (III) or cobaltic	Co^{3+}
Copper	Copper (I) or cuprous	Cu^{1+}
Copper	Copper (II) or cupric	Cu^{2+}
Iron	Iron (II) or fer-rous	Fe^{2+}
Iron	Iron (III) or fer-ric	Fe^{3+}

Common Polyatomic Ions			
1–	Name	2–	Name
$C_2H_3O_2^{1-}$	Acetate	CO_3^{2-}	Carbon-ate
N_3^{1-}	Azide	CrO_4^{2-}	Chromate
BrO_3^{1-}	Bromate	$Cr_2O_7^{2-}$	Dichro-mate
ClO_3^{1-}	Chlorate	HPO_4^{2-}	Hydrogen phos-phate
CN^{1-}	Cyanide	MoO_4^{2-}	Molyb-date
CHO_2^{1-}	Formate	$C_2O_4^{2-}$	Oxalate
HCO_3^{1-}	Hydrogen carbon-ate [bicarbonate]	O_2^{2-}	Peroxide
$H2PO_4^{1-}$	Dihydrogen phos-phate	SeO_4^{2-}	Selenate
HSO_4^{1-}	Hydrogen sulfate [bisulfate]	SiO_3^{2-}	Silicate
HSO_3^{1-}	Hydrogen sulfite [bisulfite]	SO_4^{2-}	Sulfate
OH^{1-}	Hydroxide	SO_3^{2-}	Sulfite

Common Multivalenced Cations		
Element	Name	Symbol
Lead	Lead (II) or plumbous	Pb^{2+}
Lead	Lead (IV) or plumbic	Pb^{4+}
Mercury	Mercury (I) or mercurous	Hg_2^{2+}
Mercury	Mercury (II) or mercuric	Hg^{2+}
Nickel	Nickel (II) or nickelous	Ni^{2+}
Nickel	Nickel (IV) or nickelic	Ni^{4+}
Tin	Tin (II) or stannous	Sn^{2+}
Tin	Tin (IV) or stannic	Sn^{4+}

Source: Michael O'Donnell

Common Polyatomic Ions			
1−	Name	2−	Name
ClO^{1-}	Hypochlorite	$S_2O_3^{2-}$	Thiosulfate
IO_3^{1-}	Iodate	**3⁻**	**Name**
NO_3^{1-}	Nitrate	AsO_3^{3-}	Arsenate
NO_2^{1-}	Nitrite	$C_6H_5O_7^{3-}$	Citrate
ClO_4^{1-}	Perchlorate	PO_4^{3-}	Phosphate
IO_4^{1-}	Periodate	PO_3^{3-}	Phosphite
MnO_4^{1-}	Permanganate	**1⁺**	**Name**
SCN^{1-}	Thiocyanate	NH_4^{1+}	Ammonium
VO_3^{1-}	Vanadate		

Vapor Pressure of Water											
T°C	P (torr)	T°C	P (torr)	T°C	P (torr)	T°C	P (torr)	T°C	P (torr)	T°C	P (torr)
0	4.6	17	14.5	35	42.2	53	107.2	71	243.9	89	506.1
1	4.9	18	15.5	36	44.6	54	112.5	72	254.6	90	525.8
2	5.3	19	16.5	37	47.1	55	118.0	73	265.7	91	546.1
3	5.7	20	17.5	38	49.7	56	123.8	74	277.2	92	567.0
4	6.1	21	18.7	39	52.4	57	129.8	75	289.1	93	588.6
5	6.5	22	19.8	40	55.3	58	136.1	76	301.4	94	611.0
6	7.0	23	21.1	41	58.3	59	142.6	77	314.1	95	634.0
7	7.5	24	22.4	42	61.5	60	149.4	78	327.3	96	658.0
8	8.1	25	23.8	43	64.8	61	156.4	79	341.0	97	682.0
9	8.6	26	25.2	44	68.3	62	163.8	80	355.1	98	707.3
10	9.2	27	26.7	45	71.9	63	171.4	81	369.7	99	733.2
11	9.8	28	28.4	46	75.7	64	179.3	82	384.9	100	658.0
12	10.5	29	30.0	47	79.6	65	187.5	83	400.6		
13	11.2	30	31.8	48	83.7	66	196.1	84	416.8		
14	12.0	31	33.7	49	88.0	67	205.0	85	433.6		
15	12.8	32	35.7	50	92.5	68	214.2	86	450.9		
16	13.6	33	37.7	51	97.2	69	223.7	87	468.7		
		34	39.9	52	102.1	70	233.7	88	487.1		

Source: Michael O'Donnell from various Internet sources

Functional Groups

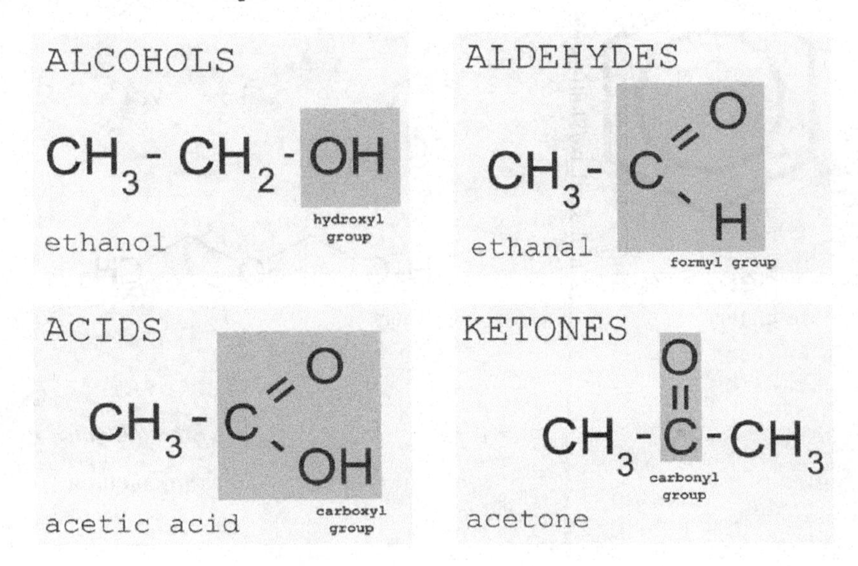

Alkenes (ethene), Alkanes (ethene), and Alkynes (ethyne)

Aromatics

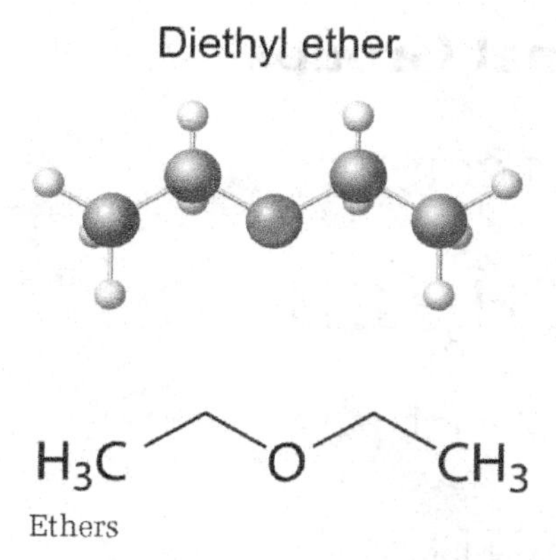

Ethers

Esters

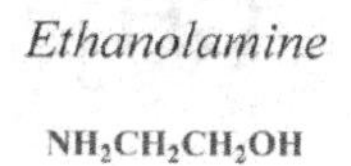

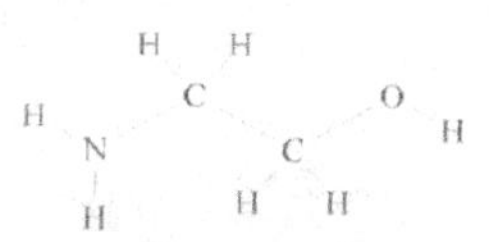

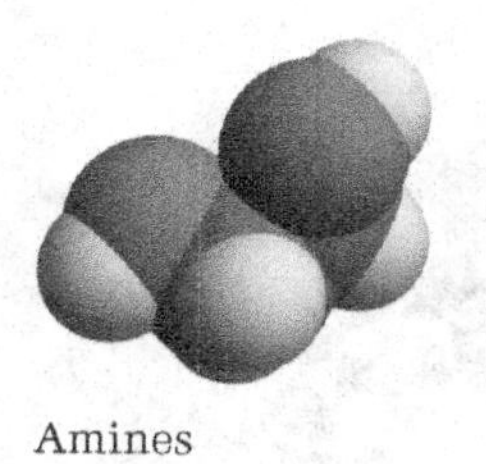

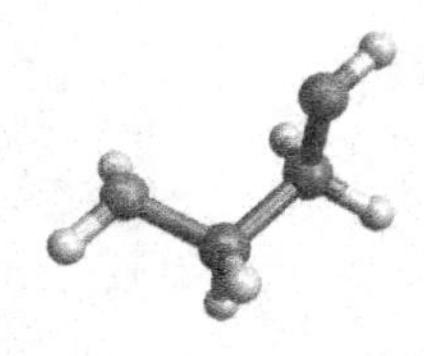

Amines

O
H_2N NH_2
urea

Amides